WASTEWATER COLLECTION SYSTEMS MANAGEMENT

WEF Manual of Practice No. 7

Seventh Edition

Edited by Cullen Carlson, PE; Bryan Jann, PE;
Mazen Kawasmi, PE, PACP, CFM; Robert Otey, PE, PACP; and
Steven Rhodes, PE, PACP

2021

Water Environment Federation
601 Wythe Street
Alexandria, VA 22314–1994 USA
http://www.wef.org

ISBN 978-1-57278-408-6

About WEF

The Water Environment Federation (WEF) is a not-for-profit technical and educational organization of 35,000 individual members and 75 affiliated Member Associations representing water quality professionals around the world. Since 1928, WEF and its members have protected public health and the environment. As a global water sector leader, our mission is to connect water professionals; enrich the expertise of water professionals; increase the awareness of the impact and value of water; and provide a platform for water sector innovation. To learn more, visit www.wef.org.

Seventh edition edited by Cullen Carlson, PE; Bryan Jann, PE; Mazen Kawasmi, PE, PACP, CFM; Robert Otey, PE, PACP; and Steven Rhodes, PE, PACP

With contributions by:

Richard Campbell
Matt Cartwright, PE
Aaron Conine, PE
Mark Evans
John Fishburne, PE
Stephen Johnson, PE
Jeff Kirkwood
Eric Nease, PE
Ryan Opgenorth, PE
Dan Purschwitz
Jeremy Rice, LEED
Matthew Schorsch, PE
Trey Shanks, CFM
Emily Tall
Jennifer Wasinger

Review provided by:

Abraham Araya, PhD, PMP
Patrick Beane
Phillip Brenner
Chein-Chi Chang, PhD, PE
Steve Donovan
Kevin Draganchuk, PE, BCEE
Brent Erickson
Joan B. Hawley, PE, PO
Michael Jankowski, PE
Jane McLamarrah, PhD, PE, ENV SP
William E. McMillin Jr., PE, D.WRE
Susan Moisio, PE
Daniel Murray, PE, BCEE
Ned W. Paschke, PE, BCEE
Chris Ranck, PE, BCEE, D.WRE, ENV SP
Diane Ryu
Michel Wanna
Christina Willson

Text originally prepared by the **Wastewater Collection Systems Management Task Force** of the **Water Environment Federation**

S. Wayne Miles, PE, BCEE, *Chair*

Abraham Araya, PhD
Nick Arhontes, PE, BSCE
Hal Balthrop, PE
Samantha E. Bartow
Robert Beringer, PE
Andy Bowen, PE
James N. Broome, PE
Gregory R. Chol, PE
Bruce J. Corwin, PE
Rich Cunningham
Ted DeBoda, PE
Robert Decker, PE
James M. Faccone
Rudy Fernandez, PE
Brent Freeman, PE
William W. S. Gray
David Guhin, PE
Rob Horvat
Richard E. Huffman, PE
James Joyce, PE
Jennifer Kauffman
Roger W. Lehman, PE
Robert Lockridge
Carol Malesky
John M. Mastracchio, PE, CFA
Robert L. Matthews, PE, BCEE
Susan R. McHugh
Amy Taylor-McMillian
Jack Moyer
Sean P. Murphy, PE
Aaron K. Nelson, PE
Paul Pinault, PE
Alison Ratliff
Ross E. Schlobohm, PE
Kelly D. Shephard
Marsha W. Slaughter, PE
Peter L. Stump, PE
Richard Thomasson
Philip Tiewater, PE
Stephen Tilson
Larry G. Tolby
Marc P. Walch, PE
Tina Wolff

Under the Direction of the **Collection Systems Subcommittee** of the **Technical Practice Committee**

Manuals of Practice of the Water Environment Federation

The WEF Technical Practice Committee (formerly the Committee on Sewage and Industrial Wastes Practice of the Federation of Sewage and Industrial Wastes Associations) was created by the Federation Board of Control on October 11, 1941. The primary function of the Committee is to originate and produce, through appropriate subcommittees, special publications dealing with technical aspects of the broad interests of the Federation. These publications are intended to provide background information through a review of technical practices and detailed procedures that research and experience have shown to be functional and practical.

Contents

List of Figures

List of Tables

Preface

For many local governments, the wastewater collection system is one of the most valuable assets they own. In addition to being a valuable physical asset, the value wastewater collection systems provide to protect public health and the environment is tremendous. Too often, however, this value is under-recognized and under-appreciated by the community. This can lead to inadequate reinvestment in the system and insufficient resources to provide for proper management, operations, and maintenance of the system.

To address these challenges, wastewater collection system managers must be exceptionally proficient in a multitude of technical and nontechnical skills needed to operate and maintain a collection system efficiently and effectively. Wastewater Collection Systems Management (MOP 7; seventh edition) provides guidance, strategies, and ideas that may be used by a wastewater collection system manager to be successful within these many fields of expertise.

The regulatory focus on wastewater collection systems has increased across the United States in the last 20 years. This is evidenced by the issuance of numerous enforcement actions by the U.S. Environmental Protection Agency (U.S. EPA) and state water-quality agencies against wastewater utilities, mandating increased system investment and program improvements. In addition, in 2005, the U.S. EPA developed the capacity, management, operations, and maintenance (CMOM) program guidelines that outline best practices for wastewater collection system managers to use to improve overall system performance. These guidelines are referenced in numerous chapters of this manual and offer a good basis for assessing whether a wastewater collection system program is comprehensive and sustainable. Although CMOM is only a guideline at the time of this writing, many state agencies that provide wastewater collection system funding and enforcement have required conformance with the CMOM guidance criteria. This is partly because of the U.S. EPA's continued encouragement to integrate management programs, like CMOM and asset management, through the National Pollutant Discharge Elimination System (NPDES) permit requirements. As such, this manual has been edited to reflect a more focused approach to essential compliance and conformance elements of maintenance plans for wastewater system managers.

With the ever-increasing changes in technology and the way people use it to communicate, many collection system owners have incorporated new channels of communicating with the public. These new channels include all the recognizable social media platforms (Facebook, Twitter, etc.) that have a broad reach, but also more localized platforms, such as neighborhood applications like Neighborhood. With the near universal use of smart phones, another common method of communicating with the public is through email or text message alerts. These new ways of communicating with the public provide the collection system manager with additional avenues to reach the public. This type of communication will not replace door knockers or formal notices at City Hall, but it will provide additional opportunities to relay important information, such as dates for upcoming public meetings, notice of service disruption, and others.

The seventh edition of this Manual of Practice was edited under the direction of Cullen Carlson, PE; Bryan Jann, PE; Mazen Kawasmi, PE, PACP, CFM; Robert Otey, PE, PACP; and Steven Rhodes, PE, PACP.

Authors' and reviewers' efforts were supported by the following organizations:

Black & Veatch

CEA Engineers

Drummond Carpenter, PLLC

Freese and Nichols, Atlanta, Georgia; Charlotte, North Carolina; Raleigh, North Carolina; Oklahoma City, Oklahoma; Tulsa, Oklahoma; Austin, Texas; Dallas, Texas; Denton, Texas; Fort Worth, Texas; San Antonio, Texas; San Marcos, Texas

Jacobs Engineering Group, Inc.

King County Department of Natural Resources and Parks/Wastewater Treatment Division

Macon Water Authority

Stantec Consulting Services, Inc.

U.S. EPA

Zhejiang University of Water Resources and Electrical Power

1

Introduction

1.0 PURPOSE

The purpose of this Manual of Practice (MOP) No. 7 (7th edition) is to provide guidance and direction to those individuals responsible for the management and oversight of wastewater collection systems. This manual can also provide valuable background information for all collection systems operators. In most urban and suburban communities, the wastewater collection system—which includes gravity pipelines, tunnels, storage facilities, lift stations, force mains, and other wastewater conveyance related facilities—is one of the largest, most valuable infrastructure assets. This system plays a vital role in the economic, social, health, and environmental viability of the community. It is important that the community recognizes the value of this system so that sufficient resources are dedicated to management, capital improvements, operations, and maintenance.

Managing a wastewater collection system poses many challenges. Aging and deteriorating infrastructure decreases system capacity and reliability while increasing operations and maintenance costs and risk of failure. Changes in wastewater flows, either from growth or a loss of significant customers, can make capacity needs difficult to manage. Third-party damage to the system and customer wastewater flows (both legal and illicit) can affect the system in ways that are difficult to predict.

A wastewater collection systems manager faces many technical, financial, social, legal, and environmental challenges. Ever-changing costs for energy, materials, and labor challenge capital and operations budgets. A lack of understanding by most customers of how a wastewater collection system works results in confusion related to how revenues are spent and how this relates to service expectations. Managers also must plan for intensified regulatory scrutiny and possible risk of third-party litigation.

To be successful, a wastewater collection systems manager requires knowledge of business, engineering, operations, maintenance, human resource management, and communications. This manual is intended to act as a reference guide on these many, varied aspects of managing combined and separated wastewater/stormwater collection systems; though it is not a complete body of knowledge on any of these topics. Additional references and lists of further readings are provided at the end of each chapter.

2.0 TARGET AUDIENCE

Although this MOP is directed primarily at wastewater collection systems managers and superintendents, many others will benefit from it. Utility engineers, system operators, local elected officials, municipal administrators, financial managers, municipal communications professionals, regulatory officials, consultants, or anyone with an interest in protecting public health and the environment will find it useful. Increasingly, the mainstream media is covering water and wastewater issues, which is helping to improve the public's understanding of the importance of infrastructure. Utility managers can and should leverage these opportunities to engage stakeholders in conversations about the importance of good wastewater system management and the value that it provides to the community.

Many cities are investing heavily in urban redevelopment projects aimed at reigniting the economic and social viability of their downtown business districts. These renewal projects offer an opportunity to engage local officials and business leaders in a discussion about the importance of rehabilitating the wastewater collection system to meet urban redevelopment and sustainability goals. Wastewater collection systems managers and personnel must be active in the planning and implementation of these renewal projects to ensure long-term success because system rehabilitation and replacement will be vital to meeting performance objectives cost effectively.

3.0 ASSET MANAGEMENT AND CAPACITY, MANAGEMENT, OPERATIONS, AND MAINTENANCE GUIDING PRINCIPLES

The U.S. Environmental Protection Agency (U.S. EPA) has been a leading advocate in the application of asset-management principles in managing our country's wastewater infrastructure (Albee, 2005). Applying asset-management principles requires utility managers to make decisions based on life-cycle cost analyses, risk assessments using condition and criticality factors, and infrastructure asset valuation effects, among others. As of the publication of this manual, no national rules have been formally promulgated requiring the implementation of capacity, management, operations, and maintenance (CMOM) programs. Many utilities, however, are adopting these programs because of state regulation requirements, consent decree requirements, permit requirements, or because of the improved system performance that can be achieved by their implementation. The principles of asset management and those promoted by CMOM programs are consistent with one another, as described further in the following sections. Although CMOM has not been formally promulgated, it is important to note that the wastewater collection system is a significant component of a wastewater treatment facility as defined by the federal Clean Water Act, and thus is regulated at the federal level through the issuance of National Pollutant Discharge Elimination System (NPDES) permits.

3.1 Asset-Management Principles

This manual of practice is focused on the management of the critical infrastructure assets that constitute a wastewater collection system. In many ways, this manual is about asset management for the specific group of assets contained in a wastewater collection system. This perspective helps to demonstrate the many areas of expertise that must be considered in the development of an asset-management program.

Managers responsible for the planning, design, construction, operation, or financing of a wastewater collection system are already managing assets. But managing assets is not the same as having an asset-management program. Managing assets includes the complex decision-making process that the collection systems manager must use every day to set priorities, consider alternatives, and make the most efficient use of available resources to accomplish program objectives. An asset-management program is a set of procedures and protocols that institutionalize, inform, document, and preserve that decision-making process. To preserve institutional knowledge and lessons learned, the importance of documenting and preserving system knowledge and experience is crucial. Although many wastewater collection systems have similarities, site-specific differences between systems must be clearly documented and understood. An asset-management program has several benefits:

- Identifies and mitigates risk and consequence of failure;
- Makes the decision-making process understandable and transparent to others;
- Provides consistent criteria for making those decisions and balancing competing needs and interests;
- Minimizes the long-term costs of system operations and maintenance;
- Defines acceptable levels of service to the customers;
- Creates data and information processes that improve future decision-making; and
- Establishes roles, goals, and metrics that can focus and motivate the entire organization toward a more cost-effective operation.

An effective asset-management program will help determine the funding needed to meet the service expectations of stakeholders, which comprise customers, elected officials, regulatory officials, the environmental community, and any other entities having a stake in the performance of the wastewater collection system. Although identifying and documenting service expectations can be difficult, not understanding those expectations can lead to longer-term difficulties. If customer level of service expectations are not met because of frequent system failures or lack of adequate performance, customer complaints may lead to undesirable political attention or, in some cases, regulatory enforcement actions. By exceeding customer expectations, the utility may increase the cost of service to greater than needed, also leading to complaints. The key is to inform the customer to increase their understanding that the level of service is directly proportional to cost. Whether the wastewater collection systems program is funded through rates, taxes, connection fees, grants, or any combination thereof, an asset-management program will provide local decision-makers with sound, understandable data illustrating the revenue that is needed to provide a given level of service and the consequences of not getting this revenue on a timely basis.

Of the many definitions of asset management, the wastewater industry has adopted the following: "An integrated set of processes to minimize the life-cycle costs of infrastructure assets, at an acceptable level of risk, while continuously delivering established levels of service" (National Association of Clean Water Agencies [NACWA] et al., 2007). The Federal Highway Administration and the American Association of State Highway and Transportation Officials (1996) provide another relevant definition: "A systematic process of maintaining, upgrading, and operating physical assets cost-effectively. It combines engineering principles with sound business practices and economic theory, and it provides tools to facilitate a more organized, logical approach to decision making. Thus, asset management provides a framework for handling both short-range and long-range planning."

Customer service expectations set the basis of asset management. Regulatory standards or local political interests typically establish a minimum level of service. The asset-management system provides a means to protect, maintain, or improve the asset value of a collection system with planned maintenance and renewal based on predicted deterioration. In a private or public utility, key asset information is needed to manage costs through asset-management planning, which requires a complete inventory of all assets. This information includes asset condition, age, and performance; operating costs; projected capital improvement costs; financial position including revenues, balance sheet, and cash flow; required and anticipated future levels of service; and methods of measuring and monitoring performance of the system.

The primary goals of wastewater collection systems operation are to prevent public health hazards and to protect the environment by transporting wastewater uninterrupted from its source to the treatment facility. The goal of asset management is to improve and document the decision-making process necessary to protect, maintain, or improve the value of this asset (the collection system) while providing the level of service desired. These issues are described further in each subsequent chapter of this manual. A glossary of terms is included in Table 1.1.

- Chapter 2—System Operations and Maintenance provides an overview of best management practices associated with operating and maintaining a wastewater collection system, including strategies to improve the efficiency and effectiveness of these operations.
- Chapter 3—Information Management describes key information management principles that are critical to wastewater collection systems management. It also describes the importance of collecting and using information to make better, more informed decisions to improve system operations.
- Chapter 4—Collection System Assessment and Capital Improvement Planning discusses the process of planning for capital improvements to the system and setting priorities, including activities needed to assess the condition of existing assets such as field activities for structural and flow evaluations and computer-based modeling tools for system capacity evaluations.

TABLE 1.1 Glossary of Terms

Term	Definition
Aerial crossing	A wastewater pipeline attached to piers to cross above a depressed area of terrain.
Cleanout	A vertical pipe with a removable cap extending from a sewer service lateral to the surface of the ground. It is used for access to the service lateral for inspection and maintenance.
Collector sewer	Sewers, approximately 200 mm (8 in.) or smaller that provide service to a localized area.
Force main	A pressurized line that conveys wastewater from a lift station to a downstream gravity main.
Gravity or main lines	The largest portion of the wastewater collection system. They use changes in elevation to transport wastewater by gravity between points.
I/I	Inflow and infiltration is the total quantity of water from inflow, infiltration, and rainfall-dependent inflow and infiltration, without distinguishing the source.
Infiltration	The introduction of groundwater to a wastewater system through cracks, pipe joints, manholes, or other system defects.
Inflow	The introduction of extraneous water to a wastewater system by direct or inadvertent connections with stormwater infrastructure, such as gutters and roof drains, uncapped cleanouts, and cross-connections with storm drains. Dry weather and wet-weather surface runoff entering the system through holes in manhole covers or other component will also contribute to inflow.
Interceptor	Large gravity wastewater pipeline pipes typically having no direct service connections that convey wastewater from trunk sewers to the treatment facility. These were typically constructed at the same time as wastewater treatment facilities, thus were designed to intercept wastewater previously flowing directly into receiving waters and convey these flows to the facility.
Inverted siphon	The portion of a wastewater pipeline that is below the overall pipeline grade as it passes under a localized vertical conflict (waterway, other utility feature, etc.). The depressed portion of the pipeline is typical surcharged in all flow conditions, resulting in the potential of debris build up over time.
Level of service	A measure by which utility operators and managers measure the quality and reliability of service or care provided to the customer based on infrastructure design standards, operating procedures, and maintenance frequency.

(*continued*)

TABLE 1.1 Glossary of Terms (*Continued*)

Term	Definition
Lift station	A mechanical method of conveying wastewater to higher elevations. A lift station is intended primarily to lift the wastewater to a higher elevation so it can continue to flow by gravity through wastewater pipeline. It can be constructed within a reasonable depth from the surface.
Lift station firm capacity	Maximum amount of wastewater flow pumped by a lift station with the largest pump out of service.
Manhole or junction box	A manhole or junction box provides a connection point for gravity lines, service laterals, or force mains and as an access point for maintenance and repair activities.
Pressure-reducing pumping station	Inline pumping station, typically variable speed and found in a manifold, force-main type of wastewater transmission system designed to reduce pressure upstream of the station.
R value	Used in flow monitoring data analysis or in a hydrologic model to represent the fraction of rainfall in a basin that enters the sewer system as rainfall-derived infiltration and inflow (RDI/I).
RDI/I	Rain-derived (or dependent) infiltration and inflow. It is I/I that occurs as a result of rain events and does not account for groundwater infiltration.
Sanitary sewer overflow	An overflow, spill, or release of wastewater from the wastewater collection and treatment system, including all unpermitted discharges, overflows, and spills. May also refer to releases of wastewater that may not have reached the waters of the United States or, in some interpretations, includes building backups.
Sewer service laterals (or service lines)	A wastewater conveyance pipe that extends from the collector system, potentially through public right of way, to the residential or other dwelling unit, or nonresidential structure to which wastewater service is or has been, provided. If the service line extends to the dwelling unit through public right of way, the portion of the line within private property is typically owned by the customer and may also be referred to as a private lateral. The portion of the line within the public right of way to the collector line is typically owned by the utility. The ownership of all or a portion of the sewer service lateral can vary from utility to utility.
Surcharge condition	Condition that exists when the supply of wastewater is greater than the capacity of the pipes to carry it and the surface of the wastewater in manholes rises to an elevation above the top of the pipe and wastewater is under pressure or head, rather than at atmospheric pressure.

(*continued*)

TABLE 1.1 Glossary of Terms (*Continued*)

Term	Definition
Transmission capacity	The capacity of lift stations and force mains that convey flow to the collection system or treatment facilities. This may also refer to the capacity for gravity pipelines to convey flow before reaching surcharge conditions.
Trunk main	Pipelines, typically larger than 200 mm (8 in.), that convey wastewater flow from the collector system to large lift stations or the wastewater treatment facility.
Tunnel	Large gravity wastewater pipeline typically having no direct service connections that convey wastewater from trunk sewers to the treatment facility. Similar to an interceptor but constructed by trenchless methods and typically deeper below ground.
Vacuum wastewater system	A wastewater system that relies on differential pressure between atmospheric pressure and a partial vacuum applied to the piping network to convey wastewater flows. This differential pressure allows a central vacuum station to collect the wastewater of several thousand individual homes, depending on terrain and the local situation. Vacuum sewers take advantage of available natural slope in the terrain and are most economical in flat sandy soils with high groundwater.

Note. Definitions or terminology used in this manual may differ from those used in local municipal codes or state and federal regulations. Legal definitions of these terms should take precedence when applicable.

- Chapter 5—System Design Considerations introduces key design principles and considerations critical to wastewater collection system managers so that life-cycle costs are adequately considered and design decisions consider risk management.
- Chapter 6—Construction Contracting provides the collection system manager with an overview of construction contracting concepts including procurement considerations, legal considerations, and how to manage quality control during construction activities.
- Chapter 7—Public Policy and Community Relations deals with ways to improve communications with collection system customers and other stakeholders with interests in collection system management as well as the development and implementation of public policy and level-of-service decisions.
- Chapter 8—Budgeting and Financial Planning summarizes key financial considerations for the collection system manager including the basis for generating revenues needed to operate and maintain the system and financial management principles needed to properly account for system depreciation and renewal requirements.

- Chapter 9—Safety, Standard Procedures, Training, and Certifications describes critical processes needed for a safe operating environment for utility employees and customers and summarizes training and certification programs that will improve understanding and compliance.
- Chapter 10—Emergency Preparedness and Security reviews issues that must be considered to prepare for both manmade and natural emergencies and disasters, including activities that can increase awareness and improve responsiveness of utility personnel during an emergency.

In addition to wastewater pipelines, lift stations, and force mains, several other assets must be managed for effective system performance. Examples include human resources that must be managed through proper training, education, and planning and information resources that require data management, quality control, and archiving standards. Other assets include intangibles such as consumer confidence, management integrity, and leadership vision. In many ways, management of these assets is as important as management of physical assets to the long-term viability and success of wastewater collection system programs.

3.2 Capacity, Management, Operations, and Maintenance Program Principles

An excellent example of the industry's increased focus on asset management is the U.S. EPA's development of guidance documents describing the key wastewater collection system principles associated with capacity, management, operations, and maintenance, also known as CMOM programs (U.S. EPA, 2005). The premise of the CMOM program is that when a utility incorporates good business principles into its organization, its wastewater collection system will meet the intended performance objectives resulting in fewer sanitary sewer overflows (SSOs). The CMOM program places the burden of proof on the system owner to demonstrate that SSOs are being prevented to the maximum extent practical by using pipes, pumps, and infrastructure with adequate capacity and by properly managing, operating, and maintaining the system. The CMOM programs promoted by U.S. EPA are consistent with basic asset-management principles as evidenced by the following descriptions of the key CMOM elements:

Capacity program elements promote an understanding of the capacity of system components (wastewater pipelines, lift stations, force mains) as designed and under current conditions, tracking how much of that capacity is used, and forecasting when additional capacity will be needed. All of these elements are necessary to support better capital improvements planning. This includes the use of flow monitoring and other evaluation tools to identify where and when capacity is needed to handle current or future increases in wastewater flows and what facilities or system improvements may best meet those needs. In some cases, these analyses may include identifying where I/I is depleting available system capacity and where the implementation of system rehabilitation to recover that lost capacity may be cost effective.

Management program elements promote efficient and cost-effective decision-making and priority-setting through sound organizational structures, good internal and external education and communications, performance goals/objectives, and proper document control and information management. Good management principles also include having sufficient legal authority, adequate resources, and clear leadership directives to meet program objectives consistently.

Operations programs must promote efficiency and reliability through the development and implementation of standard procedures, training of staff in those procedures, the tracking of system performance through monitoring systems, and the documentation of the location and condition of system components. This includes the use of Geographical Information Systems, real-time controls of system components, and inspection and testing of new construction and rehabilitation to promote the use of high-quality materials and proper installation.

Maintenance programs are the backbone of ensuring consistent system performance through scheduled system cleaning and maintenance activities as well as emergency or reactive maintenance to unanticipated events. This includes the organization and tracking of inventories for materials, spare parts, and equipment.

As collection system managers implement asset management principles, such as CMOM, a greater understanding of the existing capacity and condition of the system can be achieved. This can allow a utility to shift from reactive to proactive activities. Additionally, implementing these principles results in the collection of data—including asset attributes, condition, and work order history—that can be used by providers of machine-learning software to develop predictive tools unique to the utility.

As mentioned previously, CMOM programs are not a national regulatory requirement at the time of this publication. However, the industry is adopting CMOM principles for several reasons. Many regulatory enforcement actions have required utilities to adopt CMOM programs as terms of compliance with consent decrees or special orders of consent. In addition, several state regulatory agencies with wastewater enforcement authority require CMOM programs through collection system or NPDES permitting requirements. Other communities are proactively adopting CMOM principles because they can improve performance of the wastewater collection system.

This manual can help utility managers understand and apply asset-management and CMOM principles to improve performance of their systems. With these programs in place, managers can become a leader in better management of infrastructure assets to improve the quality of life and environmental sustainability of communities.

4.0 REFERENCES

Albee, S. (2005, November/December). Finding a pathway for sustainable water and wastewater services. *Underground Infrastructure Management.*

Federal Highway Administration & American Association of State Highway and Transportation Officials. (1996). *Asset management: Advancing the state of the art into the 21st century through public-private dialogue.* Federal Highway Administration.

National Association of Clean Water Agencies, Association of Metropolitan Water Agencies, & Water Environment Federation. (2007). *Implementing asset management: A practical guide.* National Association of Clean Water Agencies.

U.S. Environmental Protection Agency Office of Enforcement and Compliance Assurance. (2005). *Guide for evaluating capacity, management, operations, and maintenance (CMOM) programs at sanitary sewer collection systems.* https://nepis.epa.gov/Exe/ZyPURL.cgi?Dockey=30006OW9.txt

5.0 SUGGESTED READINGS

U.S. Environmental Protection Agency Office of Water. (2004). *Report to Congress: Impacts and control of CSOs and SSOs* (EPA-833-R-04-001). https://www.epa.gov/sites/production/files/2015-10/documents/csossortc2004_full.pdf

U.S. Environmental Protection Agency Office of Water. (2018). *Smart data infrastructure for wet weather control and decision support.* https://www.epa.gov/npdes/smart-data-infrastructure-wet-weather-control-and-decision-support

Water Environment Federation. (2003). *Managing the water & wastewater utility.*

Water Environment Federation & American Society of Civil Engineers. (2008). *Existing sewer evaluation and rehabilitation* (3rd ed., WEF Manual of Practice No. FD6/ASCE Manuals and Reports on Engineering Practice No. 62). McGraw-Hill.

2

System Operations and Maintenance

1.0 INTRODUCTION

Since the previous edition of this manual of practice was published, "asset management" has become a familiar topic in boardrooms, training rooms, publications, and presentations about collection systems management. The life-cycle cost of an asset is the cost to acquire, operate, maintain, and renew the asset. In its most simplistic form, asset management is a *managerial approach* or *way of thinking* that seeks to minimize the life-cycle cost of an asset. According to the U.S. Environmental Protection Agency (U.S. EPA), asset management is a "process for maintaining a desired level of customer service at the best appropriate cost" (U.S. EPA & Flennikan, 2021). Perhaps because of the involvement of maintenance management software and geographic information systems, asset management is often narrowly considered an initiative of the information technology or business technology departments.

Operation and maintenance (O&M) programs have a crucial role in asset management plans. In addition to the obvious direct costs to operate and maintain an asset, there are two other factors driving an asset's life-cycle cost that are particularly influenced by O&M programs: useful lifetime and level of service.

Operation and maintenance programs are designed to sustain a desired level of service and extend useful life by offsetting natural degradation; design, material, and installation deficiencies; and any changes to the asset's intended function or operating environment. Once maintenance is no longer able to sustain the minimum acceptable level of service, the useful life of the asset has expired and that asset may be in need of replacement.

Operation and maintenance programs must align with level of service demanded by customers, regulatory requirements, and component capabilities. To maintain a consistent level of service over time, both the cost of maintenance (because of inflation and ever-increasing regulations) and the demand for maintenance (because of diminishing asset reliability) increases, placing added importance on the efficacy and efficiency of O&M programs to prevent reduction of the financial sustainability of the collection system. To maintain the *initial* level of service for a new asset is typically more costly than maintaining a minimum *acceptable* level of service. Properly tuning an O&M program's level of effort to deliver the desired level of service over time—reliably—is one important objective of asset management.

The assumptions and directives of an asset management plan are tested in the field by O&M programs. Field activities feed essential intelligence about actual asset conditions back into the asset management plan. Whether or not a utility has a formal asset management plan, there is a symbiotic relationship between O&M programs and capital improvement plans. Well-managed, robust maintenance programs offset some of the capital demands of a wastewater collection system.

When the cost of borrowing money for capital improvements is high (or the owner's borrowing capacity is restricted), maintenance programs become more valuable because maintenance can defer or reduce replacement and improvement capital outlays. In addition, an effective maintenance program provides a defensible decision framework for capital improvements. Regardless of the owner's borrowing cost and capacity, diligent execution of maintenance programs reveals the underlying root causes of deficiencies, and therefore thwarts the potential for wasting capital investments on improvements that are irrelevant or unnecessary to extend the useful life of an asset.

2.0 PURPOSE

Intended for wastewater collection systems managers and superintendents, this chapter:

1. Summarizes the evolution of the regulatory framework and U.S. EPA efforts to incorporate capacity, management, operations, and maintenance (CMOM) and asset management plan requirements in National Pollutant Discharge Elimination System (NPDES) permits and enforcement actions.
2. Provides an overview of the various elements of maintenance programs to maximize a collection system's life cycle, such as level of service, level of effort, performance goals and indicators, information management systems, and maintenance plans.

Operation programs are addressed in other chapters of this manual.

3.0 EVOLUTION OF THE REGULATORY FRAMEWORK FOR OPERATION AND MAINTENANCE PROGRAMS

Wastewater collection systems, perhaps the greatest life-saving invention of humanity, date to as early as 4000 BCE in Mesopotamia. In the United States, sanitary engineering and municipally controlled collection systems generally began to emerge in the mid-1800s (Schladweiler, 2021). These collection systems needed maintenance to clear blockages, make repairs, and extend the system, so maintenance personnel were employed to some degree to work on the system. The modern wastewater collection system operation and maintenance industry perhaps owes its roots to 1972 amendments to the Federal Water Pollution Control Act (FWPCA).

Section 402 of the FWPCA established the NPDES permit program administered by the U.S. EPA. NPDES permits establish discharge limits and conditions for discharges from municipal wastewater treatment facilities to waters of the United States. The FWPCA authorized individual states with their own water quality standards to operate their own permit program, provided the state standards were equal to or more stringent than federal requirements. States can impose additional permit conditions based on state statutes, regulations, and rules. Most states have U.S. EPA approval to administer NPDES permits.

In 1977, the FWPCA was renamed and the nation's water statutes—along with some important additional regulations—became the Clean Water

Act. The Clean Water Act clarified that discharging to the waters of the United States is a permit-governed privilege that can be revoked and is not a right. NPDES permits include a standard condition that requires proper operation and maintenance:

> The National Pollutant Discharge Elimination System (NPDES) program prohibits discharges of pollutants from any point source into the nation's waters except as authorized under an NPDES permit. EPA and state NPDES inspectors evaluate collection systems and treatment plants to determine compliance with permit conditions including proper O&M. Among others, these permit conditions are based on regulation in 40 CFR 122.41(e): "The permittee shall at all times properly operate and maintain all facilities and systems of treatment and control (and related appurtenances) which are installed or used by the permittee to achieve compliance with the conditions of this permit." (U.S. EPA, 2005, pp. 1–5)

The U.S. EPA further emphasizes that operational failures, such as sanitary sewer overflows (SSOs), can signal ineffective maintenance and trigger a permit violation:

> SSOs that reach waters of the U.S. are point source discharges. Like other point source discharges from municipal sanitary sewer systems, SSOs are prohibited unless authorized by a NPDES permit. Moreover, SSOs, including those that do not reach waters of the U.S., may be indicative of improper operation and maintenance of the sewer systems, and may violate NPDES permit conditions. (U.S. EPA, 2021)

Thus, the conditions included in the permit that governs a collection system can be used as the starting point to define that system's minimum acceptable level of service (i.e., regulatory compliance). While most permits include performance-based standards—such as zero SSOs—some permits and many enforcement actions may prescribe specific tasks that the collection system operator must perform. (Note that some states distinguish between the collection system owner and the collection system operator. In those cases, the operator is typically a state-certified employee of the owner or state-certified contract or third-party operator, and the operator may be referred to as the "operator in responsible charge" or similar designation.) For example, the regulator may include language describing the type and frequency of inspections, or an amount of cleaning that must be performed on a prescribed schedule. In that respect, the regulatory agency may explicitly define certain components of a maintenance or operation program.

Whether the permit has prescriptive, performance-based, or a blend of prescriptive and performance criteria, collection systems managers and superintendents use regulatory criteria and customer demands to set their absolute minimum expectations (levels of service) for O&M programs. The collection system manager or superintendent then uses those expectations to design and implement appropriate programs and activities.

To help collection systems owners and operators provide a satisfactory level of service and reduce regulatory noncompliance, the U.S. EPA developed the CMOM framework (U.S. EPA, 2005).

4.0 THE CAPACITY, MANAGEMENT, OPERATION, AND MAINTENANCE FRAMEWORK

In the mid-1990s, U.S. EPA took earnest steps to close an NPDES permitting gap; that effort would lead to a proposal for CMOM regulation. The proposal to include CMOM requirements as a standard condition of NPDES permits for wastewater collection systems was primarily an outcome of the agency's objective to reduce discharges to the waters of the United States (WOTUS).

During various evaluations of wastewater utilities, the agency noticed "that good O&M relied heavily on good management" (U.S. EPA SSO Federal Advisory Committee, 1999a). The U.S. EPA had noticed that NPDES permits were issued to wastewater treatment facilities, and those operators had understood the requirements of the permits, but many collection systems were maintained by a different department that did not have knowledge of the permit requirements or understand the applicability of the permit to collection system operation. The agency also noticed that the management of information was one of utilities' greatest insufficiencies. Maintenance tools and technologies were improving organically at the time, and the deficiencies in stewardship the agency noticed were primarily attributable to administration and control efficacy. The U.S. EPA believed a "CMOM regulation" would address several areas of concern it had, including management, information systems, recordkeeping, and public notification (U.S. EPA SSO Federal Advisory Committee, 1999a).

In 1995, the U.S. EPA assembled the Urban Wet Weather Flows Advisory Committee, and the SSO Subcommittee was formed to address matters specifically related to overflows. The SSO Subcommittee evaluated permitting, enforcement, O&M principles, public notifications, and other public policy matters related to SSOs. The Water Environment Federation (WEF), along with other stakeholders, were represented on the SSO Subcommittee (U.S. EPA SSO Federal Advisory Committee, 1999b).

In November 1999, the SSO Subcommittee recommended including CMOM programs in NPDES standard permit conditions. The subcommittee further recommended including reporting, recordkeeping, and notification requirements, and prohibiting SSOs (U.S. EPA SSO Federal Advisory Committee, 1999c). Subsequently, the U.S. EPA prepared a proposal for NPDES standard conditions to improve management, emphasize system planning, and enhance public notices. The proposed regulation included CMOM requirements for municipal wastewater collection systems. In January 2001, the U.S. EPA withdrew the Notice of Proposed Rulemaking before it could be published in the Federal Register (Federal Register, 2010).

Nonetheless, the U.S. EPA continued its push to encourage holistic control of collection systems through integrated CMOM programs. In 2005, in the continued absence of an explicit regulatory tool to require CMOM programs as a condition of permit compliance, the U.S. EPA published the *Guide for Evaluating CMOM Programs at Sanitary Sewer Collection Systems*. The publication described how federal and state reviewers could assess whether a utility demonstrated compelling collection system management.

The U.S. EPA solicited input in 2010 and held a workshop in 2011 to again consider modifications to NPDES standard permit conditions for SSO reporting and CMOM requirements. As the agency contemplated how to codify the CMOM approach in proposed permit revisions, it required utilities to develop and submit CMOM programs as part of a remedy to enforcement actions. Those enforcement actions generally required the permittee to develop a CMOM program in accordance with the 2005 *Guide for Evaluating CMOM Programs at Sanitary Sewer Collection Systems*.

In 2012, the U.S. EPA provided an opportunity for municipalities to use an "Integrated Municipal Stormwater and Wastewater Planning Approach Framework" as a management tool for municipalities having to satisfy more than one Clean Water Act requirement (U.S. EPA, 2012). Even without clear-cut CMOM regulations, the agency has been able to sustain its efforts to encourage utilities to adopt an integrated management approach to collection systems O&M for the previous two decades.

Most recently, the 2019 Water Infrastructure Improvement Act amended Section 402 of the Clean Water Act to define an integrated plan as "a plan developed in accordance with the Integrated Municipal Stormwater and Wastewater Planning Approach Framework" (Gutierrez, 2019).

5.0 THE EMERGENCE OF ASSET MANAGEMENT

Alongside its efforts to encourage utilities to adopt CMOM programs, the U.S. EPA also highlighted asset management as a path to sustainable wastewater systems. A 2002 SSO fact sheet, *Asset Management for Sewer Collection Systems*, promoted the implementation of CMOM programs within an asset management framework (U.S. EPA Office of Wastewater Management, 2002). In late 2014, U.S. EPA Region 9 advocated for NPDES permitting authorities to incorporate asset management plan provisions into NPDES permits, and some regions had already implemented permits with requirements to develop components of an asset management plan (U.S. EPA Region 9, 2014).

Best management practices (BMPs) can be included as special permit conditions. In the case of requiring BMPs for the maintenance, management, and operation of a wastewater collection system, 40 CFR 122.44(k)(4) states that "each NPDES permit shall include best management practices (BMPs) to control or abate the discharge of pollutants when the practices are reasonably necessary to carry out the purposes and intent of the CWA" (Hair, 2021). Generally, the most common approach is to require the permittee to develop a BMP plan with elements such as maintenance, inspection, records, training, reporting, and risk assessment, among others. To the extent that asset management planning components qualify as BMPs, the U.S. EPA and the states have the authority to include elements of asset management plans in NPDES permit conditions.

6.0 WASTEWATER COLLECTION SYSTEM OPERATION AND MAINTENANCE PROGRAM COMPONENTS

6.1 Level of Service Definition

Operation and maintenance programs should be designed to operate and maintain a collection system to a predetermined level of service (LOS). The desired LOS is a clearly defined objective for managers to align all decisions, activities, and programs. The following example may be helpful when thinking about a simple, desired *level of service* definition for wastewater collection system as:

> We aspire to serve 25,000 residential, 35 food service, and 5 industrial contributors and convey 3.5 MGD [millions of gallons per day] of average daily dry weather flow without unauthorized discharges or odor complaints to one treatment plant without exceeding the treatment plant's 7.0 MGD peak flow capacity. We want to have personnel on the site of a reported problem within 2 hours and stop 95% of unauthorized discharges within 4 hours of arrival on site and stop any unauthorized discharge within 12 hours of arrival on site. We aim to preserve continuously fishable and swimmable waters in our streams, rivers, and lakes and we seek to maintain an overall public satisfaction rating of 4.7 on a scale of 0 to 5.0.

The LOS definition could go on to call out other requirements of the NPDES permit covering the wastewater collection system as well as other local and community requirements. The level of service could be detailed, or it could be a high-level statement with the specifics addressed in detailed performance goals the utility believes are necessary to meet in order to provide the desired level of service. The LOS must never be lower than the permit requirements.

Within an asset management framework, LOS is part of a *demand analysis*. Demand analysis is a strategic evaluation of the influences on the community's demand for wastewater collection system services. Demand analysis considers current and future demands on the system and typically includes a LOS that is the same or higher in the future. It is unusual for a utility to foresee a lower LOS for its collection system in the future, such as might occur when one or more large industries move out of a locality.

Level of service ambiguities complicate asset management objectives and regulatory compliance. When an NPDES permitting authority requires the permittee to define the desired LOS for its wastewater system, the definition must, at a minimum, describe what is necessary to maintain compliance with the permit requirements. Having simple, consistent, and quantitative measures to define LOS is important to effectively manage O&M programs, as well as any management and capacity assurance programs.

6.2 Performance Indicators

How well the system satisfies the desired LOS is typically described by a set of performance indicators that describe how often the system fell below performance standards. The indicators are used to measure the effectiveness of O&M programs. Typical "high-level" performance indicators, which might also appear in NPDES permit conditions, are

- Number of SSOs per year. Typically reported per 100 miles of gravity collection system or as individual occurrences; utilities may further designate overflows that reach the waters of the United States (WOTUS), any discharges from the collection system, or as otherwise used by the NPDES permit for reporting.
- Number of wastewater backups on private property, or annual dollar value of damage claims. Although many municipalities only maintain laterals within the public right-of-way and many minor backups are caused by private plumbing issues, significant or ongoing backups on private property can indicate excessive rainfall-dependent inflow and infiltration (RDI/I) and/or downstream main line blockages.

- Number of odor complaints. Wastewater odors can indicate hydraulic or ventilation deficiencies or can be a direct result of ongoing or previous spills. Odors can also indicate a malfunction of lift station settings. Odors can also result from temperature inversions between the collection systems and the atmosphere, which may require temporary interventions or public communications.
- Occurrences of excessive peaking factors at treatment facilities. High peaking factors reveal a system lacking sufficient containment and isolation from the environment. A poorly performing collection system can lead to treatment bypasses or damage to treatment processes. This measure reflects the effectiveness of management to address treatment and collection systems operation as an integrated system.
- Number of lift station alarms or failures. Performance measures can be further subdivided into the number of wet-weather and dry-weather failures, and how many resulted in an overflow.
- Number of swim or similar advisories annually. When clear-cut, programmatic practices exist to determine when such advisories are issued, they can be a useful LOS measure.
- Average response time. The satisfactory time to respond to public reports or requests about potential collection system deficiencies with "boots on the ground" is an important component of the LOS for a municipality.
- Average return-to-service time. Systems have somewhat unique emergency preparedness and response needs. This may have two components: the time it takes to stop a discharge from a broken pipe or overflowing manhole, and the time required to complete a permanent repair.
- Public satisfaction reports. Surveys are designed to transform qualitative data about an individual's perception into a quantitative measure of a population's satisfaction. A professionally designed, consistent survey program can provide managers with reliable data about public opinions regarding collection system management, operation, and maintenance. The number of unsolicited complaints made by the public is easily influenced by robust outreach programs, so surveys can be more reliable. Survey scores are useful measures for evaluating how well the utility's and the community's LOS definitions are aligned.

Based on a utility's specific situation, other performance measures can be used to help define the needed LOS. The NPDES permitting authority may include other special conditions in the permit that can be used as LOS measures. Most measures can be normalized (e.g., SSOs/100 miles/year) according to system mileage, number of connections, or some other appropriate denominator.

6.3 Level of Service Versus Level of Effort

Sometimes, managers attempt to use collection system condition attributes, such as age, as proxies for performance or the ability of a system to deliver the desired LOS. Caution is warranted against drawing too many conclusions about system reliability, resiliency, or ability to deliver a desired LOS from age, as many old assets have delivered flawless service for many years (Fishburne & Howitt, 2011).

It is also important that the utility's management avoid substituting level of effort (LOE) measures for LOS measures. Failure occurrences make for the most reliable and consistent measure of collection system LOS over time. Unlike treatment facilities, wastewater collection systems do not have rational scales, such as effluent concentrations, to analyze system performance. Occasionally, utilities or collection system operators mistakenly substitute LOE measures, such as miles of pipe cleaned, debris removed, or number of repairs completed, for LOS measures.

Asset management plans are designed to minimize lifecycle costs generally by reducing the required level of effort to sustain a given level of service, so using effort as a measure of performance is incompatible within a strategic asset management plan framework. Using LOE to monitor a collection system's LOS can mislead stakeholders about collection system performance and could draw regulatory and public attention to management programs and performance.

6.4 Performance Goals

Because quantitative performance indicators have numerical scales, they also serve as clearly defined performance goals. For each goal, there should be an indicator, and vice-versa. The use of "SMART" goals is helpful for managers:

S: Specific
M: Measurable
A: Achievable
R: Reasonable
T: Time-constrained

Having quantitative performance indicators and measures takes care of the first two requirements for SMART goals. Good interim performance goals might be set as annual comparisons to previous periods, which satisfies the "T" requirement for time-constrained goals. What is achievable and what is reasonable is left for debate among the managers. If the ultimate performance goal is no SSOs, but last year there were 10.5 SSOs/100 miles, it may not be achievable or realistic to set 0.0 SSOs/100 miles as a goal this year. Typically, improvement plans are developed with the NPDES regulator as a permit renewal or enforcement requirement when permittees are out of compliance. Year-over-year targets for SSOs would be evaluated and set by managers to conform with an approved continuous improvement plan. For example, the NPDES permit authority and permittee may have agreed that, under the specific circumstances, a 20% year-over-year SSO reduction goal for each of the next five years is attainable and reasonable.

There can be many more performance goals than appear in the LOS definition, because the defined LOS is the unified outcome from a collection of program and policy elements. Each of those elements should have at least one quantitative indicator that is proportional to progress toward the performance goal for the element.

Performance indicators and goals become granular over time in well-developed O&M programs. The "keep it simple" moniker is an important attribute for setting SMART goals and indicators, especially when beginning a purposeful continuous improvement effort. Information and data that is useful in a mature asset management or CMOM program can overwhelm and obstruct information management systems. The more granular the goal, the more strain is put on the organization to keep up with data and information.

6.5 Information Management Systems

In its July 1999 meeting, the SSO Federal Advisory Committee to the U.S. EPA noted in its summary comments that "EPA Region 4 looked at 80,000 work orders and lift station inspection records in the past five years and found that one of the greatest utility deficiencies is management of information. Considerable information was collected by workers in the field, but little of it was presented to management" (U.S. EPA SSO Federal Advisory Committee, 1999a).

In the 2005 *Guide for Evaluating CMOM Programs at Sanitary Sewer Collection Systems*, the U.S. EPA comments:

> Owners and operators have been shifting to computer-based systems to manage data. Only the smaller collection system owners or operators may still rely on paper management systems. Computer-based Maintenance Management Systems (CMMSs) are designed to manage the data needed to track the collection system's O&M performance. Geographic Information Systems (GIS) are used to map

and locate facilities and because of computer-based compatibility, can often easily be integrated with a CMMS. The computer-based system, however, can only be as accurate as the data used to develop it, which was most likely paper files.

By the turn of the century, CMMSs had become the standard for storing and managing system information. In the 2020s, as CMOM and asset management programs become more and more prevalent, workers probably collect much more information than they did in the 1990s. Much like looking at a dashboard crowded with instruments, the challenge for managers and superintendents is deciding which information is relevant to the decision at hand.

The SSO Federal Advisory Committee observed the difficulty utilities had managing information as geographic information (management) systems (GIS) were being introduced. GIS was a quantum leap for the management of infrastructure systems. Once management understood the power of GIS to analyze data about assets and systems of assets, and meaningfully present decision-support information, workers tended to become overwhelmed by the detailed field information they were required to collect by newly created GIS departments.

6.6 Maintenance Plans

Level of service objectives, performance indicators and goals, and an information management system are the triumvirate of maintenance and asset management plans. If any of these three strategic elements are missing, even excellent maintenance tactics will eventually fall short and system performance will be unsustainable. Tactics are described by one or more *maintenance plans*. Frequently, there are individual maintenance plans for

- Gravity collection wastewater mains and manholes
- Gravity interceptor (or outfall, large-diameter) wastewater mains and manholes
- Siphons
- Force mains and appurtenances
- Lift stations
- Real estate, easements, and rights of way
- Vulnerable or critical infrastructure

Because collection systems exist in many sizes and levels of complexity, maintenance plans may consolidate or expand on these basic categories.

Some utilities may include service laterals. Utilities with combined systems should include stormwater system maintenance or make other appropriate adjustments.

6.7 Maintenance Plan Elements

A maintenance plan should include the following elements:

1. A description of any partitions used to divide the system into manageable components of like condition, maintenance needs, or location. Easement mains and street mains are examples. In a tourist community, the collection systems serving the tourist areas might be managed as a unique partition. Mechanical, electrical, and control systems are examples of system partitions for a lift station. Maintenance partitions are generally the same thing as a system of assets in an asset management plan.
2. A description of maintenance tasks (also known as standard maintenance procedures) performed on each system partition. Visual inspection, hydraulic jetting and debris capture, root cutting, and chemical root control are all examples for gravity systems. Air/vacuum valve inspection is an example for force mains, and pump impeller trimming is an example for a lift station. Locating and marking underground assets is a maintenance task because it prevents failures as a result of third-party damage.
3. Predominant failure mechanisms observed or predicted on the system partition. Grease accumulation, vandalism, broken pipe from external loads, or root intrusion are examples for pipelines. Vibration or cavitation is an example for lift stations, and overpressurization (e.g., surge) is an example for force mains.
4. How partitions are prioritized. For example, maintenance could be prioritized according to the frequency of historical SSOs, blockages, and backups. Prioritization could (and should) also be based on the consequence of failures as well as the risk of failure.
5. Whether maintenance intervals are time-dependent or condition-dependent for system partitions. Typically, maintenance plans with the objective of providing a consistent level of service use time-dependent maintenance on partitions in which the cost of the conditional test exceeds the cost of the maintenance, or where no reasonable conditional test exists. Time-dependent maintenance intervals are also preferred when degradation occurs at a relatively predictable rate. Condition-dependent maintenance is typically preferred

for systems or components that require expensive maintenance or may have highly variable times-to-failure.

a. For time-dependent (i.e., preventive) maintenance, describe how the optimal maintenance interval is determined.
b. For condition-dependent (i.e., predictive) maintenance, describe the conditional tests used to determine the need for each maintenance task. Like time-dependent maintenance, describe how the interval for condition testing is optimized.

6. A description of equipment and tools necessary to perform each of the prescribed maintenance tasks. Inventory levels and location of equipment and tools should also be included, as should required staffing levels.
7. A discussion of maintenance production capacity under the plan. Provide production estimates for each maintenance task; availability of mobile equipment, tools, and staff; and whether there are mutually exclusive tasks, such as the same crew being used for many maintenance tasks.
8. A maintenance schedule for tasks on each system partition based on reasonable or historic production. The ability to develop and adhere to a preventive maintenance schedule proves a utility's progress away from being controlled by reactive maintenance and validates the maturity of its strategic asset management plan.
9. A framework for improvements to the maintenance plan. The framework should address what information is collected about failures, maintenance tasks, production, observed conditions, costs, and resources used, and how that information is recorded and incorporated in information management systems, continuous improvement plans, and capital improvement plans. This might also incorporate an escalation plan for recurring maintenance that is unsustainable because of economic or other reasons, such as being irrelevant to the failure mechanism (such as routine cleaning of a pipe with a large sag).
10. Periodic revisions to goals, indicators, processes, and practices to improve the plan's effectiveness to meet LOS objectives. Generally, these revisions to the plan are the result of comprehensive evaluations that are part of a continuous improvement program. Performance goals may need to be more aggressive as the program matures. Perhaps indicators prove less relevant than expected to judge performance and need adjustment. Processes and practices improve as new skills and tools are developed.

7.0 REFERENCES

Federal Register. (2010, June 1). U.S. EPA notice. *Federal Register, 75*(104), pp. 30395–30401.

Fishburne, J., & Howitt, I. (2011). Reliability centered maintenance: Gravity main blockage condition testing and degradation progression estimation using the SL-RAT Acoustical Testing Tool. *WEF Collection Systems 2011*. Water Environment Federation.

Gutierrez, S. (2019, December 3). *Implementation of integrated planning in accordance with the 2019 Water Infrastructure Improvement Act.* U.S. Environmental Protection Agency. https://www.epa.gov/npdes/implementation-integrated-planning-accordance-2019-water-infrastructure-improvement-act

Hair, D. (2021, January 31). *Special or narrative conditions in NPDES Permits.* U.S. Environmental Protection Agency. https://www.epa.gov/sites/production/files/2021-01/documents/special_or_narrative_conditions_in_npdes_permits.pdf

Schladweiler, J. C. (2021, February 19). *Tracking down the roots of our sanitary sewers.* Sewer History.org. http://www.sewerhistory.org/time-lines/tracking-down-the-roots-of-our-sanitary-sewers/

U.S. Environmental Protection Agency. (2005). *Guide for evaluating CMOM programs at sanitary sewer collection systems.*

U.S. Environmental Protection Agency. (2012, June 5). Integrated municipal stormwater and wastewater planning approach framework. https://www.epa.gov/sites/production/files/2015-10/documents/integrated_planning_framework.pdf

U.S. Environmental Protection Agency. (2021, February 19). *National Pollutant Discharge Elimination System (NPDES): Municipal wastewater.* https://www.epa.gov/npdes/municipal-wastewater

U.S. Environmental Protection Agency Office of Wastewater Management. (2002, April). *SSO fact sheet: Asset management for sewer collection systems.* https://www.epa.gov/sites/production/files/2015-10/documents/assetmanagement.pdf

U.S. Environmental Protection Agency Region 9. (2014, December). *Asset management: Incorporating asset management planning into NPDES permits.* https://www.epa.gov/sites/production/files/2018-01/documents/incorporating-asset-mgmnt-through-permits.pdf

U.S. Environmental Protection Agency SSO Federal Advisory Committee. (1999a). *Meeting summary comments, July 29, 1999.*

U.S. Environmental Protection Agency. SSO Federal Advisory Committee. (1999b). *Meeting summary, October 20, 1999*. U.S. Environmental Protection Agency.

U.S. Environmental Protection Agency SSO Federal Advisory Committee. (1999c). *Consensus committee recommendation*. U.S. Environmental Protection Agency.

U.S. Environmental Protection Agency & Flennikan, A. (2021, January 21). *Asset management 101: Getting started with asset management webinar handout*. https://www.epa.gov/sites/production/files/2021-02/documents/asset_management_101_webinar_handout.pdf

8.0 SUGGESTED READINGS

New England Interstate Water Pollution Control Commission. (2003). *Optimizing operation, maintenance, and rehabilitation of sanitary sewer collection systems*. https://www3.epa.gov/npdes/pubs/sso_optimizing_enitre_doc.pdf

Water Environment Federation. (2010, July). *Core attributes of effectively managed wastewater collection systems*. https://www.wef.org/globalassets/assets-wef/3---resources/topics/a-n/collection-systems/technical-resources/2010-08-02fca-1.pdf

3

Information Management

1.0 INTRODUCTION

This chapter focuses on the information management needs and requirements of the collection system manager. Presented here are general management principles—rather than complex technical elements—based on the assumption that technical and supervisory resources are available elsewhere to implement the necessary software and hardware solutions.

Three forces have arisen that are compelling collection system managers to evolve away from paper-based (or non-existent) recordkeeping to computer-based systems: federal and state environmental regulators, such as U.S. Environmental Protection Agency (U.S. EPA); asset-management requirements (Government Accounting Standards Board [GASB] 34); and overall improvements to the industry (technical certification programs). In theory, paper-based systems can fulfill the analysis and reporting requirements of these three demand areas, but the cost of manually collecting and reporting the data is not financially acceptable in light of the availability of contemporary digital systems.

However, the evolution toward digitalization has not been without its problems. For example, digitalization can create a flood of raw data from multiple sources that can result in an incoherent picture, thus becoming a source of frustration and difficulty—rather than assistance—for a collection system manager. However, when properly filtered and sorted, the data can become knowledge, which when combined with a collection system manager's experience, can yield insight, leading to effective resource decisions. The objective of this chapter is to provide the reader with a sufficient framework with which to make those decisions in a balanced and professional manner.

2.0 ESTABLISHING AN INFORMATION MANAGEMENT SYSTEM

Collection systems have historically been viewed by the public and elected officials as low-tech and somewhat simple in nature. Therefore, establishing a digital information management system (IMS) can be more challenging for the collection system manager to execute than say, for example, the purchase of a wastewater vac truck, which is a much more routine and familiar proposition. An IMS can consist of one or more software packages featuring geographic information systems (GIS), computerized maintenance management systems (CMMS), data analytics, and data visualization capabilities. An IMS requires a noticeable investment to establish and maintain, including the cost of staff labor, and to a lesser degree, software and hardware. Such an investment will be a new, added, and permanent cost for the utility, and these factors tend to draw the scrutiny of elected governing boards. However, in many cases the IMS for the collection system can be a module within an enterprise software package that is utilized by other entities within the organization.

To make the establishment of an IMS as palatable as possible to governing boards, the manager needs to develop a case based on the highest possible goals and rationales rather than on just the technical merits. Emphasizing the overall public benefit of an IMS will engender the broadest political and community support, compared with the more limited success that might be achieved if the merits are described using complex, difficult-to-understand computer-based terminology.

To assist the manager in developing a rationale for a governing board, the following are the most common and defensible principles upon which to establish a collection system IMS:

- Conform to regulatory responsibilities, principally U.S. EPA's capacity, management, operations, and maintenance (CMOM) program and its variants;
- Perform timely and consistent information searches;
- Maintain recurring work orders scheduled for preventive maintenance and inspections;
- Provide justification for operations budgets;
- Track customer calls as well as historical maintenance and repair work orders;
- Organize capital rehabilitation and replacement plans based on asset-management data;
- Preserve the utility's institutional knowledge;
- Provide real-time, user-friendly information to field staff; and
- Provide useful and timely maintenance trend and productivity reports.

2.1 Conformance to Regulatory Responsibilities

Since the mid-1990s, collection system agencies have been subject to varying degrees of regulatory requirements embodied in CMOM. An IMS is an integral part of CMOM because of the accuracy and consistency that can be achieved with well-structured and well-executed computerized management programs.

CMOM programs, as described in the U.S. EPA *Guide for Evaluating Capacity, Management, Operation, and Maintenance (CMOM) Programs at Sanitary Sewer Collection Systems*, incorporate many of the standard operation and maintenance activities that are routinely implemented by utilities (U.S. EPA, 2005). It does, however, create a new set of information management requirements, including

- Provide timely and relevant information;
- Better manage, operate, and maintain collection systems;
- Investigate capacity constrained areas of the collection system;

- Proactively prevent sanitary sewer overflows (SSOs); and
- Respond to SSO events.

The IMS requirement has been incorporated by numerous state and regional regulatory bodies into their own versions of CMOM, and these requirement(s) can be relied upon to form the development and use of an IMS within the collection system utility.

2.2 Performing Timely and Consistent Information Searches

The speed with which information can be retrieved from the Internet has raised the public's and media's expectations how local government delivers data. This has obvious implications for collection system managers who are subject to the same expectations for speed. It makes the processing power of an IMS an invaluable asset that collection system agencies cannot do without because poring over reams of paper records to retrieve, collate, and distribute information is a grinding, error-laden task that does not speak to the highest possible professionalism of a collection system utility. Outlined below are the most frequent demands for information retrieval.

- Customer and media inquiries are almost always related to complaints or problems; otherwise, there is almost no interest on the part of the public in the normal operations of a collection system. The ability to provide an immediate reply using information retrieved electronically leads to high marks awarded for customer service.
- Elected officials expect to be kept informed on the performance of the collection system and its engineering and operations staff, especially during rate-increase requests. A well-organized database can yield compact, uniform reports that bolster the confidence of elected officials and provide a sound basis for making resource decisions.
- The GASB promulgated Rule 34 to measure the degree to which the infrastructure of the United States is being maintained. It seeks to establish a profile of the life span of the nation's infrastructure and when it will need to be replaced or rehabilitated. Although GASB is not an infrastructure management organization, it is responsible for setting the standards by which the financial health of the cities and agencies is measured because they must be able to establish the revenue needed to keep their infrastructures intact and functioning. An IMS designed to yield current and projected infrastructure data can be used to satisfy GASB 34 reporting requirements. Engineering data,

asset-management data, inspection records, and maintenance histories form the foundation of GASB 34 reporting.

- When a collection system utility is faced with a damage or liability claim, gathering and arranging historical data is difficult without an IMS. Attorneys who represent agencies in litigation are at a distinct disadvantage when records and reports are not digitally organized. In addition, a utility can depend on its IMS to alert them to repetitive conditions such as flooding patterns that can leave them exposed to damage claims.

2.3 Maintaining Recurring Work Orders for Preventive Maintenance and Inspections

The repetitive nature of preventive maintenance and inspections can create a mind-numbing experience when trying to organize, assign, and report activities without an IMS. Using an IMS to keep track of commitments made over long periods of time helps to ensure reliability and continuity. In fact, preventive maintenance and inspections that have long intervals between activities are the ones most likely to be forgotten without the functionality to assign recurring work orders, which is built into an IMS. Paper reminders or other manual scheduling methods are more likely than not to become lost or forgotten, especially for long-interval work.

The GASB 34 standard has specific timetables for inspection intervals, which can be loaded into the IMS. Because pipe inspections tend to occur at long intervals, an IMS becomes invaluable for this task.

2.4 Providing Justification for Operational Budgets

Operational budgets are typically calculated on the basis of resources needed for each of four areas: preventive maintenance, inspections, complaint servicing, and spot repairs. Without an IMS, many collection system managers are left to define their budgets based on historical experience or intuition, neither of which has solid professional footing.

An IMS that is properly loaded with data can easily create a profile of the correct budget by examining resource requirements that are defined for each of the four areas of operations. For example, a manager can use the IMS to quickly and accurately sum the total labor hours required for a year's worth of preventive maintenance to determine if the budget is sufficient to accomplish all the tasks. If the manager determines that there are 10,000 labor hours of preventive maintenance work to do in a year, but has a budget for only 8000 labor hours, it is easy to see that the utility has an accumulating deficit of work, also known as "backlog," and they will never "catch up." These types of calculations are readily available in an IMS and can provide validity to elected officials for the need of an adequate foundational budget.

Additionally, an agency that can never keep up with its preventive maintenance schedule could face negative consequences from regulators who would be intolerant of an inherently deficient preventive maintenance program.

2.5 Tracking Prioritized Work Order and Repair Schedules

Prioritization of work orders is a dynamic process because collection systems themselves are hydraulically and structurally dynamic. In addition, environmental factors such as wet weather add to the process. Because an IMS operates in real time, the priority shifts that will occur, also can be reported in real time. This provides an accurate picture of the demands on a utility at any given moment. This is especially important during emergencies when a manager will be called upon to explain the service status of the organization. An IMS-based report quickly will be able to display both the specifics and the global status of the situation. These reports can be displayed as dashboards with real-time information.

Additionally, linking the status reports of customer calls from the IMS to the organization's website, provides the broadest information distribution to the public. This goes far in reassuring the public that the service they are expecting is not lost in an indeterminate "black hole" of local government. An IMS-based report can also be used to satisfy a regulatory audit to verify priority work orders are being properly fulfilled.

2.6 Organizing Capital Rehabilitation and Replacement Plans Based on Asset-Management Data

Prioritized capital replacement and rehabilitation plans that are linked to asset-management information are similar to prioritized work order and repair schedules as described above. However the sheer scale and expense of doing capital work requires a separate prioritization scheme that is much more complex than the relatively simple variables associated with work orders.

Several collection system attributes can cause asset management to be complex: structural materials and performance; maintenance and repair history; anticipated future flow; the ability of the agency to finance capital work; and much more. Capturing this information and producing a rational capital management plan based on this degree of complexity is only achievable with an IMS as the cost of assembling and analyzing the information manually is often prohibitive.

The rise of sophisticated asset management as a refined science is due primarily to the availability of data analysis capabilities associated with computer systems. Until recently, the complexity of this information has been too overwhelming to cost-effectively analyze using other means.

2.7 Preserving the Utility's Institutional Knowledge

Of all the principles upon which to establish and maintain a collection system IMS, preservation of institutional knowledge is perhaps the most important of all. A collection system is a long-lived public asset and the knowledge of how the system was built, operated, maintained, and rehabilitated/replaced is a knowledgebase that will outlive current and future utility staff.

Too often, the retirement of a knowledgeable employee has meant that a substantial part of the utility's institutional knowledge was lost, to the detriment of those still working there and to the ratepayers who support the system. For this reason alone, an IMS that can maintain this knowledge across generations of employees is a valuable asset for a manager.

3.0 BUILDING AN INFORMATION MANAGEMENT SYSTEM

Once the premise and rationale for an IMS has been established and approved, the process of building it can begin. This process involves an understanding of the types of data associated with collection systems, the attributes that need to be captured, and the reports that need to be generated.

3.1 Types of Collection System Information

Building IMS for a collection system involves an understanding of the different types of data found in a typical system. In analyzing the information management needs of a utility, the manager should be familiar with the basic types of data that are typically associated with collection systems.

Collection system information can be categorized into five principal types:

1. *System* data is the heart of collection system information, upon which all other data are dependent. System attributes consist of an inventory of a collection system's assets, its maintenance, repair, inspection, and condition assessment. Customer service records and the history

of complaint resolution are also components of system data. (Failure to maintain customer service data may invite litigation as a result of claims for damages.)

2. *Spatial* data describes the location of system structures with respect to real world geographical references, typically survey coordinates, and often are derived from global positioning satellite readings. Spatial data is the backbone of GIS, which are used to create electronic maps of collection systems and link data to them.
3. *Dynamic* data is the reporting of real-time data that describes how the collection system is functioning. The most common dynamic data is lift station activity and alarms reported by supervisory control and data acquisition systems; flow metering results; and rain gauge readings.
4. *Modeling* data is used to determine the existing and future hydraulic capacity of a collection system. It is derived from sophisticated computer software that melds inventory parameters, such as pipe size, with topographic data, such as slope, with flow metering results and storm activity to profile the system's ability to handle flow during dry-weather and wet-weather events.
5. *Asset value* data is the calculation of the monetary value of the collection system, the cost to rehabilitate or replace it over its lifetime, and the revenue demands necessary to support those capital improvements. There is a close link between how the system is maintained and the length of its life expectancy.

3.2 Selecting the System Attributes to Retain in a Collection System Information Management System

A collection system manager should aim to have a set of criteria by which to judge whether or not the utility should expend resources to capture, store, and report an attribute. The goal should be to gather a sufficient, but not overwhelming, amount of information to allow a reliable understanding of the collection system. The five basic collection system attribute groups are listed at the end of this section.

It is important that before the information system is built, the manager defines the ongoing daily, weekly, monthly, and yearly reports that will be required from the IMS. These output reports will dictate the input requirements of the IMS, which will eliminate gathering and inputting information of questionable value. The manager should start with the end in mind, because waiting until the system has been implemented to try and design ongoing reports will be expensive to correct.

Also, there must be an appropriate balance between the capability of the software and the capability of the agency using it. The sophistication of the desired software should not exceed the staff's ability to use it, as this could result in a significant loss of investment. Conversely, agencies need to closely analyze the promised capability of proposed software installations to ensure that it will actually deliver the desired functions.

To help facilitate the attribute selection process, a collection system manager should ask several questions about each.

- Does it identify a unique value, such as a structure's identification or location, or a customer? Unique values are the most important attributes to capture because databases and maps depend on this data to navigate finding, sorting, and filtering data.
- Is the software user friendly? Software that is difficult to use can result in slow uptake by the organization and clerical errors.
- Does the software offer mobile (tablet, phone, laptop, etc.) capabilities? Increasingly IMS software packages can be deployed at the crew level, providing critical information to staff working in the field.
- Does the software integrate with the organization's existing IMS software packages? In many cases, an organization's IMS consists of multiple software packages (GIS; CMMS; SCADA Historian; Data Analytics, Reporting, and Visualization; etc.), so understanding the compatibility with these legacy software packages is critical before purchasing and deploying new software.

Statistical analysis and reporting are the primary products of an IMS because they exploit the calculating speed of computers. Incidental information that would not be subject to statistical analysis may not be worth the effort. For example, is it worth the effort to know the size of each manhole cover in a system?

There are several good examples of statistical data worth keeping and reporting:

- A daily list of incomplete work orders,
- A summary of spent materials to manage operational budgets,
- A sum of the length of all the pipe in the system,
- The quantity of assets inspected over a given period,

- The number of SSOs per approximately 160 km (100 mi) of pipe in the system, and
- The number of service laterals in the system.

Other questions include: Is the attribute a problem flag? Does the attribute serve to alert you to an important condition that is otherwise hard to remember? Does it warn you of an effect or potential effect on public health and safety, or on the safety of crews?

For example,

- Does a structure have an exceptionally hazardous aspect, such as a nearby industrial waste discharger?
- Is it a work-order location that is known to be barricaded and, therefore, more hazardous to the public?

Does it inform the manager of important contractual or ownership issues regarding a wastewater structure?

For example,

- Is it a structure that is privately owned or maintained, and is, therefore, not the utility's responsibility?
- Is a structure's construction guarantee still in effect when a repair is required?

Is the attribute spatial-dependent data, and is it needed for building and maintaining the collection system maps or hydraulic model?

Examples of map or hydraulic model data are:

- Latitude/longitude survey data,
- Rim elevation,
- Invert elevation,
- Depth, and
- Length.

If the attribute is an inspection defect, is it properly coded, formatted, and documented such that it can be incorporated into the IMS?

Inspection defects are the single most abundant attribute in a collection system IMS. They are typically derived from television inspection records, and with inspection equipment it is possible to capture endless amounts of defect observations. A significant effort to standardize the observation, coding, digital storage format, and documentation of inspection defects has been undertaken by the industry. Many organizations have adopted the standards developed by the National Association of Sewer Service Companies and Pipeline Assessment Certification Program for pipelines, Manhole Assessment Certification Program for manholes, and Lateral Assessment Certification Program for service laterals. All inspection defects and associated condition data should be stored in the IMS to provide a historical perspective.

However, it is not possible or realistically desirable to repair every observed defect because the cost would not be justifiable to perform trivial repairs. Collection systems, like all infrastructures, are in a constant state of change and the most cost-effective maintenance and repair is that which addresses significant problems, not all problems.

There are five basic collection system attribute groups that would comprise both an IMS and a GIS:

1. Structure data
   - Manholes;
   - Main lines;
   - Service laterals;
   - Catchbasins or stormwater inlets;
   - Lift stations, comminutors, and generators;
   - Septic tanks and miscellaneous structures; and
   - Stormwater storage/detention facilities.
2. Maintenance data
   - Scheduled preventive maintenance;
   - Scheduled repair work orders;
   - Unscheduled/emergency work orders; and
   - Weekly, monthly, and yearly system maintenance and crew productivity reports.
3. Inspection data
   - Structural defects
     - Pipeline (including service lateral) defects-cracking, deflections, etc.;
     - Pipe joint defects-offsets, etc.; and
     - Manhole and catchbasin/stormwater inlet defects-structural components.

- Maintenance conditions
  - Sags
  - Debris
  - Grease
  - Roots
  - Odor
  - Vermin
- Inflow and infiltration measurement

4. Hydraulic/capacity defects-surcharging, overflow frequency; only a subset of the factors needed to determine if level of service or regulatory requirements are met.
5. Condition assessment—a combination of specific physical conditions (defects) and structural data (material type)

3.3 Practical Guidelines for Establishing an Information Management System

One of the most important mantras to remember is: garbage in, garbage out. Installing a computer system cannot repair the errors of a poorly managed, paper-based system. If a utility's paper records are in disarray, mere data-entry of those records will only exacerbate the problem of trying to remedy the errors. Additionally, when establishing an IMS or changing software packages, it must be determined if historical information is in a format that is valuable and worth the time to incorporate into the new IMS software. For example, if work orders are addressed based on no reference to a physical asset, it would be difficult to relate those work orders to an asset in the IMS. Although spatial techniques can be applied to assume a relationship to a physical asset based on proximity, this effort would only be an approximation. It may be better to store this information outside the IMS for future analysis and start a collection of asset-centric data with the establishment of the new IMS. Like house painting, the majority of the work is in the preparation, not the final step.

The biggest cost of computerization is data collection, input, and upkeep—not software and hardware. When a utility undertakes the establishment (and operation) of an IMS, there is a tendency to underestimate the labor hours required to build and use the system, including staff training. Those labor hours, which are a real cost, cannot by trivialized and should be a primary focus when considering an IMS. The cost of the IMS software and hardware, even some of the most expensive systems, does not even compare to the required labor commitment. Incorporating data management and integrity practices, such as using available regional data (county parcels, city limit boundaries, waterways, etc.), minimizing free form input, automating report development, and providing refresher training, can help mitigate labor expense of maintaining the IMS.

4.0 OPERATING AND MAINTAINING AN INFORMATION MANAGEMENT SYSTEM

Operating and maintaining the IMS is vendor-specific and should be addressed by its supplier. However, in larger, more complex organizations, there is a benefit to having a dedicated, in-house expert who will maintain the platform and work with vendors and employees to keep the system up to date and functioning. There is one aspect of the IMS that only the collection system manager can supply: security of the data against accidental loss because of fire, employee mistakes, vandalism, or cyberattacks.

Every labor hour that is put into building and using the IMS increases the economic worth of the system in replacement-value terms; the IMS could eventually become the utility's single most valuable asset. Information stored in the IMS is contained in one or more databases, depending on the number of software packages used. An established procedure to regularly back up the IMS database(s) should be developed. This backup copy of the production IMS database(s) should be stored on a separate piece of hardware (cloud environment, server, computer, portable hard drive, etc.) such that it can be quickly put into production with minimum data loss in the event the production IMS database is lost. Additionally, as with all components of an organization's digital infrastructure, strong security protocols should be implemented to prevent hacking and ransomware attacks. This includes strict user password requirements which must be changed at some frequency (quarterly, bi-yearly, yearly, etc.).

5.0 REFERENCE

U.S. Environmental Protection Agency. (2005). *Guide for evaluating capacity, management, operation, and maintenance (CMOM) programs at sanitary sewer collection systems* (EPA 305-B-05-002). https://nepis.epa.gov/Exe/ZyPURL.cgi?Dockey=30006OW9.txt

4

Collection System Assessment and Capital Improvement Planning

1.0 INTRODUCTION

A well-documented capital improvement plan (or program) (CIP) provides a utility with a sound understanding of the costs and schedules of system improvements that will be required to sustain performance in the future. An accurate CIP allows managers and decision makers to understand near- and long-term financial needs, which is critical to sustaining a consistent level of service at a consistent cost to customers.

This chapter will outline the process of developing and updating a CIP and describe the information required to conduct this process. Because CIP development requires an understanding of both existing system conditions and projected future needs, this chapter includes a discussion of system assessment approaches and technologies. Capital improvement plans are a multiyear planning instrument used to identify needed capital projects and to coordinate funding needs and timing to maximize value to the public. The key components to a successful CIP include

- Reliable tools for inspection, monitoring, and recording the condition of the collection system infrastructure;
- Methods and tools to evaluate capacity, performance, and the effects of future growth on the existing system;
- System to incorporate and prioritize input from the key stakeholder's interests including engineering, operations, finance, management, and customer;
- Priority matrix developed with input from the key stakeholders; and
- Appropriate funding levels and phasing of improvement strategies to achieve desired replacement goals.

This chapter guides the wastewater collection system manager through the important steps in developing a CIP. The CIP process includes developing funding requirements that will be used for budgeting, which is discussed in more detail in Chapter 8. In general, to develop a CIP a manager needs to follow several key steps:

1. Identification,
2. Validation,
3. Prioritization, and
4. Financing (see Chapter 8).

2.0 SYSTEM ANALYSIS

Understanding the system is the first step in developing a successful CIP. Various tools and methods are available to help the manager gauge the pulse of the collections system. These tools are vital in identification and prioritization of system improvements.

2.1 Capacity Assurance Planning

Most municipalities have been experiencing population growth and, as a result, an increase in flow to existing infrastructure. In addition, infiltration and inflow (I/I) increases as infrastructure ages; these increased flows can result in basement backups and sanitary sewer overflows. Hydraulic models and use of other analysis tools—including spreadsheets and Manning's equation—typically are used to identify size alternatives to relieve capacity restrictions for various storm and build-out conditions. Other tools such as geographic information systems (GIS), flow monitoring and supervisory control and data acquisition (SCADA) are useful in identifying potential capacity issues with an increase in growth and I/I. For an overview on I/I and how its effect on a collection system can be evaluated using hydraulic models, refer to the Water Environment Federation (WEF) fact sheet, *Sanitary Sewer Systems* (WEF, 2017).

Managers should track incremental flows added to the system, which can affect capacity. The capacity of all new developments should be recorded to eliminate the possibility of "double dipping" by allocating the same capacity to two different developers based on the timing of actual construction. This can be accomplished using GIS or a simple spreadsheet as new connection fees are collected. The data can then be used for various modeling simulations to determine the potential effect on the system.

2.2 System Analysis Tools and Methods

The process of developing a CIP has evolved over the years as new tools and methods for assessing wastewater collection systems have emerged. Tools now used for development of a successful CIP include hydraulic models, automated field data collection, asset-management systems (GIS and CMMS [computerized maintenance management systems]), and sophisticated databases. These tools, when used together, can assist the manager in making well-informed, defendable, and optimized decisions about which projects to include and what priority each receives.

2.2.1 Data Collection and Management Methods

Acquisition, analysis, and application of collection systems information is the foundation of accurate and effective decision-making in a CIP. The most effective way to collect and analyze this information is by using an asset-management system. A utility's ability to effectively manage its infrastructure is directly linked to its ability to maintain accurate and up-to-date information on the status and location of its facilities. At a minimum, this requires that changes made by field crews, such as repairs and small replacements performed through operations and maintenance (O&M) efforts, be recorded and shared with staff maintaining the facility's maps, records, and engineering information.

Information about a collection system's inventory should be linked to its geographic location to provide a spatial context of the extent and condition of the system. This link between the data and spatial component is achieved through the use of GIS. Through the use of unique identification numbers, geo-coded addresses, and global positioning system records, assets can be viewed in the correct geospatial context for a variety of uses, including locating where critical assets exist and how they are grouped.

Asset-management systems and GIS are useful tools for developing a CIP only if the data that populates these systems is accurate, reliable, and reproducible. Various data collection methods are available to achieve this goal including SCADA, global position systems, field data collectors, flow monitoring, and collection system inspection techniques (such as closed-circuit television [CCTV], smoke testing, manhole inspections, multisensor inspections, and destructive testing).

2.2.2 Supervisory Control and Data Acquisition Systems

Supervisory control and data acquisition systems provide continuous monitoring and control of remote points. In wastewater, SCADA typically is used to monitor collection system lift stations and to provide real-time alarming and historical archive of system operations for analysis and predictive maintenance. Data collected through a SCADA system provides key information for development of a CIP, including pump run times, increase in head pressure or volume of I/I flow, and a history of system deficiencies identified by real-time alarming. The SCADA helps the wastewater collection system manager identify future problems and plan and budget for future CIP projects.

2.2.3 Geographic Information Systems and Computer-Aided Design

One of the most common problems in collection systems management is determining the locations of collection system lines and manholes. This task is best done by maintaining maps showing the locations of the collection system lines, cleanouts, and manholes; depth-to-pipe invert; pipe material and size; and the direction of flow. Pumping stations, appurtenances, and water resource recovery facilities (WRRFs) should be clearly identified. With the advancement of mobile computing, maps of the collection system can be made available to field personnel. Maps also can show the location of other utilities like water, stormwater, power, natural gas, and other franchise utilities.

System mapping for many utilities consists of large, paper maps divided into overlapping, large-scale sections. These maps or sheets are bound into books that can be stored easily and taken into the field as needed. For older systems, original maps may not be accurate because construction over the years may have relocated entire sections. If not maintained properly, maps can quickly be outdated by development, reducing accuracy and confidence-level.

Developments in computer mapping have made it possible to detail systems and easily add new structures. Computer-aided design and GIS applications have expanded to offer a wider variety of uses, from simple mapping to computer-aided modeling for predictive management. Among some of these parameters exist the ability to provide

- Asset attributes (unique ID, diameter, material, etc.),
- Information layering and dimensional,
- Variety of maps for different collection system applications or designs,
- Easily revisable maps,
- Record of activity sites and causes of problems,
- Record of current and future rehabilitation sites,
- Record of current and projected maintenance sites, and
- Record of television inspection sites.

2.2.4 Computerized Maintenance Management Systems

Collection systems contain several asset types that need to be managed and maintained: mains, manholes, siphons, force mains, cleanouts, pumping stations, laterals, diversion structures, remote storage facilities, and automated valves. Some regulatory bodies that manage discharge permits are requiring agencies to have a system in place to schedule regular maintenance, inspection, and cleaning of the collection system and document these activities. As discussed in earlier chapters, asset management is a process that addresses the efficient, long-term management of the collection system from an operational engineering and economic perspective. Asset management is a sustained, systematic process of designing, operating, and maintaining assets and facilities effectively. It is important to remember that asset management is not a software system but rather a programmatic approach to managing and maintaining assets that is accepted and supported by the entire organization. Computerized maintenance management systems are software packages that can assist an agency in maintaining a computer database of information about an organization's maintenance operations. This information is intended to help maintenance workers perform their jobs more effectively and to inform management decisions. Today, most systems are computer-based and easy to use. Advances in computer technology have made the development and maintenance of databases cost effective for even the smallest utilities. Systems should be tailored to meet unique facility needs. A careful evaluation of specific needs, based on clear objectives, should guide development of an effective system. A discussion of information management systems can be found in Chapter 3, Section 2.0.

Specific categories of information necessary for computerized maintenance systems are facilities inventory, service requests and work orders, and maintenance history. Inventory data should be available within the CMMS so that jobs can easily be planned and crews can be sent to locations with the appropriate parts and materials. Data entry into the CMMS system should be entered through the domain list when possible to normalize the data and identify trends. This information is important to the organization for other uses, such as asset management, capital planning, or hydraulic modeling. Historically, location and attribute data have been maintained on maps and drawings, inventory cards, or manual records. Therefore, if digital records are not kept, the data conversion activity will be the largest effort in the development of a CMMS.

2.2.4.1 Maintenance History

Maintenance history data are valuable for determining budgets, planning capital improvements, and developing preventive and predictive maintenance activities. Maintaining records serve two purposes: to help record work previously performed in the system and to help plan future projects. Records of the maintenance activities are typically tracked on prepared forms corresponding with a particular CMMS format. These forms include information such as the date, location, work performed, time to complete the work, distance, reason for work, material removed, equipment used, condition of assets, and any additional remarks.

Accurate and current maintenance history can also protect the agency in case of a litigation challenge. During litigation, lawyers for both sides typically ask for maintenance history of specific items within certain dates. The CMMS must be capable of handling queries used to obtain the required information without too much manual input or data recreation.

Utilities currently using paper systems should consider computerized record keeping because of its lower long-term cost and ability to process information in seconds instead of hours or days. Using a CMMS gives a manager the ability to manipulate data in a variety of ways, from relational comparisons of different pipe materials and conditions, to automate work-order generation based on maintenance or inspections completed the day before. All are generated far faster than written records at far less cost over time. Some of these programs allow all operations (biweekly cleanings, staggered yearly maintenance, or repairs to the system or pumping stations) to be scheduled, tracked, and analyzed over time.

Once maintenance history and facility inventory data are available through GIS, the data can be viewed in conjunction with other information, such as a collection system's maintenance history overlaid with results from a hydraulic analysis. This compilation helps to prioritize capital improvement projects and to protect assets.

2.2.4.2 Complaints and Service Requests

Whenever a complaint is received from a customer, a service request needs to be recorded on either paper or in the CMMS. It is important that these requests be tracked and reviewed on a regular basis to identify trends and problem areas before they escalate. Guidelines for handling customer complaints must be developed and understood by all maintenance employees. Each agency should develop their own benchmarks to use in identifying areas for improvements. Various benchmarks have been developed for responding to complaints and other customer service issues, which are described later in this chapter. Charting customer complaints gives the manager more information for identification and prioritization of future capital projects. Service requests typically initiate a work order to address the problem if additional work is identified. A best practice is to tie the service request to an individual asset ID so the root cause of the problems can be tracked and preventative measures implemented. Additionally, a standard process should be developed to identify which work orders are routine or urgent.

2.2.4.3 Work Orders

The primary function of most CMMS applications is service order processing. All work orders should be geographically referenced, assigned to an asset, and should provide information about the assets involved. Specific assets play key roles in scheduling and tracking maintenance service orders and ultimately are included in a CIP. Work orders typically include the assets involved; the problem reported; activities performed; materials, parts, and labor required; and other job-site information. Asset and GIS facilitate retrieval and use of this information by identifying specific facilities, locations, or attributes.

2.2.5 Integrating Geographic Information Systems and Computerized Maintenance Management Systems

Many utilities have converted to computerized systems to handle schedules, plans, specifications, engineering drawings, contracts, change orders, time sheets, and project communications and management. Computerized systems allow managers more time to manage projects; better and faster communications; less time and money spent producing reports, plans, and specifications; and a reduction in clerical staff.

Computers are playing a critical role in managing a utility's spatial and tabular data as described above. The growing availability of geospatial data and GIS has given managers a valuable tool in planning, identifying trends, understanding the system, and presenting findings to staff, governing boards, and the public.

Integrating CMMS and GIS data with an existing system is one of the biggest challenges in creating a functional planning tool. This integration, however, can yield the greatest benefits. For municipalities, this includes looking at how data flows though the department, reviewing workflow processes, and potentially re-engineering the information systems that support the organization. Because approximately 75% to 80% of the organization's information relates to location, proper implementation of a GIS can dramatically improve the effectiveness of this information. Automation and integration of work orders and linkage of the CMMS with the GIS allows maintenance supervisors, engineers, and managers to statistically analyze complaints, recurring problems, and repair data. When linked to a GIS, data can be presented using custom queries to show pin maps or other thematic maps that show the type, frequency, and clustering of events. For example, a collection system's maintenance history can be overlaid with results from a hydraulic modeling analysis to rank potential capital improvement projects.

Combining the CMMS and GIS data into an integrated tool allows managers the ability to more effectively plan and identify capital improvement projects that will

- Address recurring problems,
- Comprise projects that are spatially diverse across the municipality,
- Combine similar projects in close proximity, and
- Include information on surrounding construction projects.

2.2.6 *Inflow and Infiltration Identification*

The volume of I/I in a collection system is a good indicator of overall system health and can be used to help identify and prioritize CIP projects. These extraneous flows can affect system performance, decrease capacity of existing infrastructure, and significantly increase the cost of wastewater treatment. Inflow and infiltration exist in all collection systems, and inspection crews are frequently tasked with identifying and tracking I/I parallel with condition inspection. Rehabilitation, refurbishment, and replacement of assets that are the sources of the most I/I in a system is typically a large part of the overall CIP.

Inflow is extraneous stormwater or snowmelt runoff that enters a collection system through roof leaders; cleanouts; foundation drains; sump pumps; and cellar, yard, and area drains. Stormwater also may enter through cross-connections between a sanitary sewer and storm sewers and through defective manhole covers and frame seals.

Infiltration is water that enters a collection system from the ground through defective pipes, pipe joints, damaged service connections, or manhole walls. Infiltration most often is related to high groundwater levels but can also be influenced by storm events or leaking water mains.

Rainfall-derived infiltration and inflow (RDI/I) is I/I that occurs during and following a rain event. It has been recognized as a significant source of operating problems, often contributing to sanitary sewer overflows (SSOs) and backups into homes and basements. The RDI/I also can cause serious operating problems at wastewater treatment facilities because it leads to increased flows, which can sometimes overload hydraulic or process capacity of plant facilities.

Inflow issues typically are easier to identify during wet weather because drains or sump pumps become active. These inflow sources typically are privately owned and need to be terminated by the property owner causing the discharge. In many instances, the property owners are not aware that they are discharging stormwater to the collection system.

Infiltration can be more difficult to pinpoint and more expensive to resolve than inflow. Although the amount of infiltration can be estimated with some accuracy by a proper flow-monitoring plan, the actual infiltration sources can be especially difficult to find. They consist of underground defects that admit stormwater or snowmelt that has soaked into the ground. Though some leaks can be found during and immediately after wet-weather events, many will take days or weeks to find as groundwater levels increase enough to reach the collection system. Flow monitoring typically is used to find these areas of extraneous flows, which are then televised during periods of high groundwater levels to determine where the flows originate. After being cataloged, repairs can be made on a priority schedule based on estimated volume of infiltration.

The identification and quantification of I/I flows in the system typically is done by using testing techniques such as flow monitoring or portable instantaneous flow weirs. The use of CCTV, smoke testing, and dyes can also provide information on the sources of extraneous flows. Some large utilities have found it economical to keep chemical grouting and sealing equipment on hand during CCTV activities for spot repairs. It is important to catalogue all defects identified before making repairs; data can be used in future analysis and prioritization of similar pipe for CIP projects.

Inflow and infiltration data collected during inspection are used to identify and quantify specific sources of extraneous flows and determine which segments are affected. After significant wet-weather flow monitoring and instantaneous weir measurement data recording, flows are analyzed to determine where subsequent wet-weather and dry-weather I/I investigation work should be performed. The objective of an I/I investigation is to identify extraneous flow sources so that those pipes can be targeted in the CIP.

Field investigations typically include several tasks:

- Flow monitoring and instantaneous weir measurement to:
  - Determine base wastewater and extraneous flowrates,
  - Determine collection system basins with excessive extraneous flows, and
  - Identify collection system reaches with excessive extraneous flows.
- Internal manhole inspection to identify extraneous flow sources in manholes and tributary pipes.
- Wet-weather television inspections to determine extraneous flow sources and distribution between public and private property laterals.
- Dry-weather, CCTV inspections.
- Dry-weather smoke testing.

- Dry- or wet-weather dye water testing at suspected storm sewer direct and indirect connections to the collection system.
- Wet-weather surface inspections for manhole inflow sources.
- Building inspections for the identification of illegal connections such as roof leaders, sump pumps, and foundation drains during construction and remodels.
- Identification of cross-connections such as storm water drains.
- CCTV inspections of service connections with extraneous flow or maintenance problems.
- Rainfall data collection.
- Review of historical work order data.

These investigations are necessary to identify and control extraneous flows that hydraulically overload some collection systems, pumping stations, and WRRFs. In most instances, severe extraneous flows will activate SSOs and cause backups that create health and environmental problems. Reduction of extraneous flows reduces the cost of wastewater conveyance and treatment, increases available system capacity, and reduces the need for additional collection system and treatment capacity.

Hydraulic models described later in this chapter are a valuable tool in determining the effects of I/I on system components listed above and the benefits of its elimination.

2.2.7 Flow Monitoring

Knowing how gravity flow and pumping stations convey wastewater through the collection system is essential for understanding how the collections system functions on a day-to-day basis. Knowing the depth and velocity of wastewater moving through the system and determining what the theoretical volume should be based on the number and types of users can help determine what percentage is normal flow and what is attributable to I/I.

Obtaining accurate flow measurements is useful when developing a CIP. Flow measurements can help identify areas with high rates of I/I and insufficient capacity during wet-weather events. In addition to locating extraneous flows, measurements inform the system manager of the amount of wastewater being conveyed versus the capacity and estimated user volume of the system overall or in a given area. When development begins to tax capacities of pipes, flow measurement may be the only means available to regulate land use or system upgrade parameters. This information also is valuable for bypass pumping or for sizing a new pump for a lift station. Flow measurements can be used to determine I/I by comparing dry- and wet-weather measurements. In some cases, flow measurements are used to detect wastewater leaking out of the pipe and water entering it. As opposed to totalizer flow monitors—which typically are recorded daily and provide total flow quantities between monitor readings—continuous monitoring can be used for more detailed analysis. For example, continuous monitoring can be used to measure treatment flowrates from joining communities for billing purposes, and projected extraneous flowrates and volumes. Older monitoring facilities initially used depth-only monitors to measure flowrates, which can be inaccurate if submerged flow or backwater conditions occur or if debris buildup affects the primary measuring device. Typically, these monitors were used in permanent or temporary manhole locations, providing a printed record on a circular chart on a daily or weekly basis. Advances in technology now provide monitors that can read depth and velocity simultaneously, providing greater accuracy during normal and surcharge flow conditions over longer periods of time. Data typically is stored electronically and manually retrieved from the monitor, or a telemetry system can be used to transfer information periodically or in real time.

Temporary monitoring is used to determine various conditions, including

- Dry-weather wastewater flowrates;
- Influence of extraneous flows;
- Flowrates and depths for computer modeling;
- Peak and average daily flows for lift station design, bypass pumping, system upgrade, and wastewater interceptor design;
- Available capacity for development;
- Location of bottlenecks;
- Project exfiltration rates; and
- Collection systems with maintenance or capacity difficulties.

Many utilities use flow-monitoring devices to measure wastewater flows from sub-basins in their collection systems to assess the effects of I/I. Dry-weather, wet-weather, and rainfall-induced flow conditions are recorded. The dry-weather diurnal flowrates are used to identify base wastewater discharges and the presence of groundwater infiltration. Wet-weather flow conditions occur during and after rainfall and typically lead to the highest rates of flow in the collection system. Analysis of flow-monitoring data can be used to distinguish between wastewater flow components and identify where RDI/I flowrates and volumes are the highest, thus providing priorities for collection system investigation and rehabilitation.

Projected groundwater infiltration and RDI/I rates are determined through numerical and graphical comparisons of dry- and wet-weather flows, along with seasonal variation of groundwater levels and antecedent moisture conditions. A graphical comparison of dry- and wet-weather flow patterns is an excellent method of projecting I/I rates and visualizing effects at the monitoring site. Figure 4.1 is a hydrograph (a plot of wastewater flow during a selected period of time) that projects RDI/I during a 24-hour period. A rainfall-induced flow plot is superimposed over a typical dry-weather flow pattern for the same day of the week (for example, Monday versus Monday) at the same monitoring site. The hydrograph notes the peak rate and volume of RDI/I during the analysis period. Through this type of flow comparison, the daily flow patterns or cycles at a particular monitoring site are matched. Any difference in the flow patterns is attributed to wastewater fluctuations or I/I.

Along with flow monitoring of extraneous flows, rain gauges should also be used to determine rainfall amounts, duration, and intensity. Comparisons

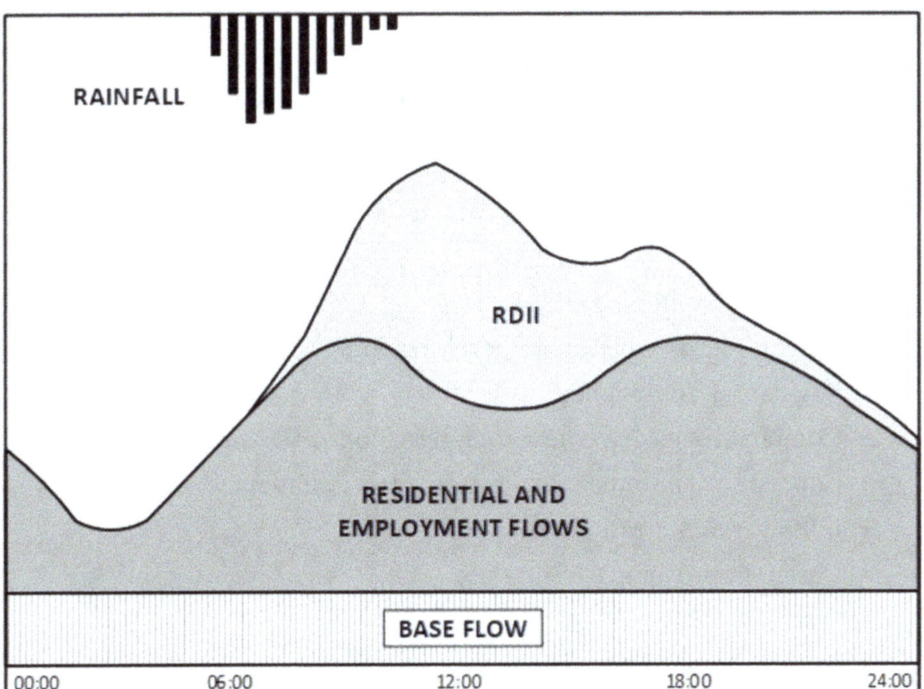

FIGURE 4.1 Hydrograph Depicting the Direct Discharge, Groundwater Infiltration, and RDI/I Over a 24-Hour Period

between rainfall and flow monitoring determine the relationship of rainfalls to collection system extraneous flow responses.

The number of flowmeters and rain gauges used can vary greatly depending on the goal of the monitoring program. Various guidance documents provide insight on potential number to be used and should be reviewed, then balanced with budgetary constraints.

2.2.8 Hydraulic and Hydrologic Modeling

Hydraulic and hydrologic modeling is a powerful tool that managers can use to evaluate the current and future state of the collection system. The material in this section provides a high-level description of the critical components of a hydraulic and hydrologic model. For a more detailed discussion on this topic, refer to *Existing Sewer Evaluation and Rehabilitation* (WEF & American Society of Civil Engineers, 2008). This section uses excerpts from that manual for purposes of giving the wastewater collection system manager enough information to evaluate the required data and effort to use a hydraulic model for use in collection system management. The utility should review established guidance on modeling protocols and standards, then adopt one that is best suited for their situation to ensure consistency in their modeling efforts.

Hydraulic and hydrologic models traditionally have been used as planning tools and to efficiently evaluate alternatives and solutions. In addition, when merged with GIS and other applications, models can extend beyond planning projects and become an integral part of an infrastructure management toolbox. In this capacity, models can serve multiple objectives:

- Master planning;
- Capital improvement plan development;
- Regulatory compliance;
- Combined sewer overflow long-term control planning;
- Capacity, management, operations, and maintenance assessments;
- Inflow and infiltration evaluation;
- Design improvements;
- System O&M;
- Real-time control modeling;
- Water quality/watershed studies; and
- Climate resiliency.

The development of a model varies depending upon the intended objective and purpose of the model and the information required from it. Most municipalities will have two primary objectives:

1. Develop a model to accurately simulate operation of the existing wastewater collection system and its ancillary structures.
2. Develop a model that will become an operational and planning tool to guide ongoing maintenance activities and assist in the allocation of future wastewater collection system capacity to new development.

For capital planning, it is important to understand the level of sophistication of the model used and what assumptions are being made in the development phase. Hydraulic models are often created for specific purposes, such as capacity assessment or regulatory requirements, but should be continuously updated to incorporate system improvements, expansion, and operational changes. Some modeling approaches may lead to recommending a more conservative approach to ensure that the system will be sized to handle the worst case scenario, but this may cost more than decision makers are willing to spend. Understanding the benefits and limitations of a model will allow managers to set an acceptable level of risk as they develop their long-term CIP.

2.2.8.1 Steady-State Simulation Versus Dynamic Simulation

When looking for a hydraulic model, it is important to have a good understanding of the system and what is important to its operation. If the system routinely surcharges and pipes are serving as storage, then the model used must accurately be able to predict the shape of the hydrographs in both surcharged and non-surcharged flow conditions. Other factors that should be taken into account are whether it is necessary to simulate:

- Surcharge levels or flooding,
- Backwater effects in nonsurcharged pipes,
- Reverse flows in nonsurcharged pipes,
- Looped systems, and
- Modeling of ancillaries (e.g., pumping stations and regulators).

Where there are complex control systems or ancillary structures, special consideration should be given to the capability of the model to simulate these features. If the model is to be used as the basis for detailed collection system modeling, this will have to be considered in selecting software.

Existing steady-state models are not sufficient to develop this level of understanding. Although they have their place for some applications (design of new systems, for example), they are not suited for developing a detailed understanding of existing system operating characteristics required by most collection system owners today. Dynamic modeling will provide

- Detailed understanding of collection system performance;
- Ability to project into the future;
- Accurate understanding of I/I mechanisms;
- Basis for prioritization of improvements;
- Optimization tool, including real-time control alternatives;
- Integration of collection and WRRF models; and
- Supporting regulatory program development.

Dynamic modeling of the collection system should be used to determine available system capacities and, from that, where the collection system can be used to provide interim capacity relief for the treatment plants. Further, dynamic modeling will allow the analysis of real-time controls for developing capacity augmentation measures that may include remote or other flow control measures and/or inter-basin transfers within a municipality's collection system.

2.2.8.2 Flow Development

For modeling purposes, it is useful to further define the components of flow in wastewater collection systems. Figure 4.1 displays a typical 2-day wastewater flow hydrograph. Flow is plotted against the *y*-axis on the left side of the plot and precipitation intensity is plotted against the inverted *y*-axis on the right side of the plot. The wastewater flow hydrograph can be broken into three basic components:

- *Groundwater infiltration* is typically measured during the overnight hours when the minimum amount of direct discharge is observed in the collection system; it may have seasonal variations but is typically treated as constant over the duration of a single precipitation event.
- *Base wastewater flow* enters the collection system through direct connections and represents the sum of domestic, commercial, and industrial flows.
- *Rainfall-derived inflow and infiltration (RDI/I)* is the portion of the wastewater flow hydrograph above the normal dry weather flow

pattern; it is the wastewater flow response to rainfall or snowmelt in a watershed. Inflow is typically represented by a rapid, sharp spike in flow, whereas infiltration occurs at a slower rate over a longer period of time. The effects from inflow may recede soon after a storm event has concluded if sufficient conveyance capacity exists, while infiltration may linger for longer periods until soil saturation is reduced or groundwater level recedes.

2.2.8.3 Calibration/Verification

Model calibration is the process of adjusting estimated model parameters to match observed flows and water depths in the system. This step is essential for the creation of a model that accurately represents a collection system. In general, there are few parameters that can and should be adjusted; the final value should always fall within the range of reasonable values for that parameter. If a value outside of that range is needed to achieve a suitable match between model results and field measurements, the model likely is not accurately representing the system. This is one of the reasons that inspections may be required: to check that actual system configuration (size, depth, elevation, etc.) matches available information. In general, calibration proceeds in the downstream direction. To accurately calibrate the hydraulic model, two efforts are needed: dry-weather calibration and wet-weather calibration. The goal of calibration is to simulate flow in the collection system within an acceptable range of observed flow data. Calibration targets also should take into consideration modeling objectives and availability/reliability of all data sources used to develop the model.

Typical tools used in the calibration and verification of hydraulic models include

- Flow monitor data (dry- and wet-weather data);
- Rain and snowfall data (point and radar);
- Facility information (i.e., lift station set points, run times, pump curves);
- Number of connections and estimated water usage;
- Land use;
- Soil type, impervious cover, topography;
- Infrastructure asset information (location, diameter, length, invert elevation, etc.); and
- Recorded overflow information.

The required level of calibration, type of calibration (such as single event, multiple events, or continuous simulations), and flow monitoring are determined after defining project goals and carefully reviewing existing information on dry- and wet-weather hydraulics. With adequate calibration, the models become consistent tools to assess the hydraulics of the existing collection system, identify changes in flowrates, and identify capacity concerns to be incorporated to a CIP.

2.3 Condition Assessment

Assessment of the collection system condition should be part of the daily activity of the inspection and maintenance staff. When determining the replacement priority of an asset, age, and material, or calculated expected useful life is a valuable piece of information, but actual condition is much more useful. Everyone involved in field activity should gather information on system conditions as they perform their assigned tasks. For example, in addition to setting up and performing flushing and cleaning activities, operations staff could perform an above-the-ground manhole inspection at the receiving end while waiting for the equipment. This type of joint activity maximizes the use of operations staff time and also provides valuable information to engineering staff for CIP prioritization and I/I analysis.

Collection systems management is charged with maintaining the system and the investment it represents. An asset inventory tracked in a CMMS can provide the reporting and inventory analysis required by management. The data, if collected and stored properly, can also be used for planning and predicting system upgrades and future capital plans. A well-managed system should track several items:

- Population served by the system;
- Total system size in length of lines;
- Inventory of pipe lengths, sizes, and access points;
- System appurtenances, bypasses, siphons, overflows, diversions, and splits;
- System component installation dates and materials;
- Total theoretical capacity of the system;
- Total theoretical generation of wastewater by population;
- Industrial and commercial users with high discharge levels;
- Total current flow volume;
- Total length off-road easements and rights-of-way;

- Force main locations and lengths;
- Number, location, and capacity of pumping stations;
- Number and location of siphons and inverted siphons;
- Number, location, and size of tide or flood gates;
- Equipment on hand (maintenance, excavating, and repair equipment);
- Number of field personnel available;
- Breakdown of pipe sizes and materials;
- Breakdown of interceptor, trunk, branch, and lateral lines maintained;
- Condition of physical line sections as available;
- Manhole inventory—location, size, type, and condition;
- Pumping station components, including pumps and valves;
- Number and locations of aerials and crossings; and
- Complaint totals, including type, location, cause, and frequency.

There are many more detailed items that can be added, depending on the size, complexity, and topography of the system.

The Water Environment Research Federation (WERF) has developed protocols for assessing the condition of wastewater utility assets (American Works Water Association [AWWA] Research Foundation & WERF, 2007). The research outlined a 10-step approach to developing a condition assessment program to align data collection and decision-support efforts:

Step 1. Document program drivers.
Step 2. Specify program objectives.
Step 3. Identify asset types to assess.
Step 4. Collate and analyze available data.
Step 5. Determine what assets to inspect, if any.
Step 6. Select inspection/assessment technique.
Step 7. Plan inspection program to minimize cost.
Step 8. Undertake asset inspection and other data collection.
Step 9. Analyze data and assess asset condition.
Step 10. Use condition assessment information for decision-making.

2.3.1 Surface Inspection

Some problems can be detected by visual inspection while walking or traveling along the rights-of-way of the collection system line, both in the street and in off-road easements. Sidewalk irregularities; cracked, settled, or dipped pavement; ponding over manholes; or depressions along the path of the pipe can be indicators of leaking or collapsing pipes. If a joint is bad or a pipe is broken, wastewater may wash away the surrounding embedment and create a cavity beneath the surface. Sometimes, the weight of the overlying soil is enough to cause collapse and a depression at the surface. In easements, these depressions can be seen as flooded or sunken areas along the pipe route, especially after a significant wet-weather event. Surface inspection can also be invaluable for critical lines that cross streamways or run parallel. Those lines can be exposed or damaged during high flow rain events, so having a planned inspection can identify those issues before they lead to an overflow event.

Cavities under the road surface are often reported as a depression in the street. Such a report may go to a street repair crew, and the depression would be filled without further investigation. To avoid this, the cooperation of the street repair department is essential. Cavities that are left unattended will enlarge, and the weight of a vehicle can be enough to cause a road collapse. Depressions (settling) can also be the result of inadequate compaction during construction or subsequent utility work.

Manhole inspections are another source of surface surveillance information. Manhole inspection technologies include pole camera, LiDAR scanning, and 360-degree panoramic photography, in addition to visual or similar methods. The National Association of Sewer Service Companies (NASSCO) has developed a national standard for assessments called the Manhole Assessment Certification Program. This program provides a standardized assessment format to note items such as evidence of high water or surcharging; evidence of grease accumulation or root intrusion; missing bricks, concrete, or rungs; and water entry through the barrel, riser, frame, or cover of the manhole. Broken, raised, or sunken manhole frames or covers can be noted and repaired quickly.

Manholes periodically are examined for any sign of bedding material in the invert that suggests a breech or collapse in the upstream line. Deteriorated steps may indicate the presence of corrosive gases, such as hydrogen sulfide. Information on the completed forms is transferred to the computer or directly entered into a mobile application and loaded to the data file to be used later for listing priority, difficulty, or common traits.

In both surface and underground inspections, evidence of water entry should be noted for future consideration or solution. Adding this information to the system inventory will create a far more comprehensive database from which hydraulic modeling can accurately simulate actual conditions and be used to predict system upgrades or repair needs.

2.3.2 Closed-Circuit Television

Some older methods of visually inspecting pipe are still used, such as manhole entry to lamp portions of collection system pipes, entry into large pipe for visual inspections, and the use of mirrors on adjustable poles. However, most internal inspection is performed by the use of CCTV systems. Televising of pipes eliminates the need of personnel entry into manholes and can provide a video record of conditions in existing pipes.

Television inspection systems operators should be trained in equipment operation, collection system defect classifications, and data recording procedures. Any defects in the pipe should be noted on the video, and the appropriate defect code should be noted in the software that typically accompanies inspection equipment. Most CCTV inspection software now has the ability to link video with inspection results, which give the user the ability to jump directly to the defect. These defect records should be documented according to the coding or rating system adopted by the investigating agency. There are numerous coding systems in use, ranging from basic, handwritten codes with 10 to 20 classifications, to computer applications that contain numerous code classifications and matrixes. NASSCO has also developed a national standard for pipeline assessments called the Pipeline Assessment Certification Program (PACP). An increasing number of collection system operators are moving to the PACP coding system, which is supported by many of the inspection software programs. The PACP data can be used to develop a condition renewal program based on prioritized risk factors established by organization and industry best practices.

Computer systems allow storage of digital video and inspection results for rapid processing and cross-referencing that can be used to evaluate and rank large quantities of data. The agency should select a coding system that is appropriate for its particular applications and train operators to properly use it to provide consistency in defect coding. Some utilities use the operator's rating system to help determine priority levels. If a rating scale is used, operators should be trained in making assessments because most ratings can be subjective to some degree.

Data management is required after surveillance records are collected. Some utilities continue to use a paper system that splits information into areas of responsibility so that it can be reported to various departments. Others are able to tie the inspection data directly into the CMMS and GIS for use in tracking costs, time, and condition of pipeline. In addition, by tying directly into a CMMS, the information collected by field operations staff can be made available throughout the utility. This sharing enables the work-order management system to issue repair requests based on severity and priority of the problem identified in the field. This procedure takes less time and allows the data to be handled only once. Priorities and repair or upgrading projects are clear so that long-term budgets can be used and crisis situations avoided.

2.3.3 Smoke Testing

When investigating for I/I in the system, the inspection crew frequently uses smoke testing as a means to identify direct inflow connections and system failures that should be repaired or addressed in a CIP. Smoke testing involves a specially mounted blower that forces air and smoke into a manhole and pipe system. The smoke generated is non-toxic, has no odor, and typically is foggy-white in color. The smoke is forced by the blower into the line and seeks the path of least resistance to exit the pipe. Under average circumstances, the structures in the section of pipe being tested will release this smoke from the vent stacks of the buildings. If gutter drains are connected to the service line, smoke will appear from the downspouts. As a precaution, when engaging in smoke testing, descriptive information should be supplied to every building on the line before the testing. Many utilities use a simple door hanger that informs the public of the work being done to lessen any concerns of toxicity of the smoke.

When conducting smoke testing operations, the fire department should also be notified if there is a sump pump or floor drain connected to the service line through which the smoke will exit the well. Also, if there are dry traps or defective plumbing fixtures in the building, smoke can enter. People could mistake the smoke for a fire. It is important to note that smoke testing typically is not effective in locating all sump pump connections. Some pump systems incorporate "P" traps or check valves that block the smoke or are in basements that are not viewable from the street or lawn areas. Additionally, defects in the lateral (roots, sags, etc.) or backflow preventers can block the transmission of smoke. If a vent stack does not smoke, that should also be noted for future reference.

Most property owners are not aware of illegal connections to the system. In such cases, when smoke is viewed exiting from anything but the vent stack, a photo should be taken and shown to the owner and attached to the inspection report for future reference. Plans can then be made to terminate the connection, typically at the owner's expense. In some areas, local, state, or county funding may be available to help defray the cost to system users of disconnecting these sources.

Smoke testing also can show cross-connections to storm sewers and other underground leak points. Smoke will exit from storm drains or appear out of the ground, indicating a problem. These problems will require visual inspection of the pipe to determine where the opening or cross-connection is and can sometimes be difficult to trace because the smoke can drift between underground pockets and appear far from the actual release point. In these cases, dye water testing can be used to trace the connectivity.

Building inspections are another means of inflow control. Though time-consuming, inspections may be the only way to find illegally connected sump pumps and floor or foundation drains. Building inspection can be conducted while smoke testing is ongoing, with few property owners not allowing access to structures. Some utilities work with the local fire and building departments when they inspect private and commercial property, asking them to note any pumps or obvious drainage that goes to the service connection.

2.3.4 Trunkline Inspection

Trunklines and large-diameter pipe typically are inspected and maintained on a separate schedule and with different equipment than smaller collection system lines. Trunklines are a critical component to a collection system in their carrying capacity and in the potential for catastrophic failure. Special attention to these pipes is important for maintaining a safe and operational collection system.

Multi-Sensor Inspection (MSI) is a method of inspection for large-diameter pipe. Features of MSI include laser to measure the diameter of the pipe above the water line. Sonar or laser measures the amount of debris deposited below the water line at the bottom of the pipe. Closed-circuit television shows the condition of the pipe. Defects like I/I, cracked or damaged pipe, and exposed rebar are assessed to aid in the decision-making process of repair, rehabilitation, or removal. In a large-diameter pipe inspection, it is important to identify the critical crossings and to prioritize based on the available funds. These sensitive crossings could include highway, railway, and water bodies. Results from these inspections should be recorded and prioritized in a CMMS and GIS for use in determining capital projects. Typically, replacement of a trunkline has a large financial effect on a municipal budget. By regularly inspecting these trunks, potential problems can be identified early and added to the CIP for rehabilitation to be budgeted in later years.

2.3.5 Pumping Stations

An asset-oriented maintenance management system can reduce the time and cost of pumping-station management. Computer-based maintenance management systems offer additional advantages. The steps required to order a replacement pump part illustrate the large amount of information that must be stored, tracked, and integrated:

- Issuing a work order,
- Obtaining equipment nameplate and master parts list data,
- Checking spare parts inventory, and
- Purchasing or replacing the part in inventory.

In addition to maintaining the facility, using a CMMS will assist the manager in planning for and funding capital improvements. A CMMS will allow the manager to track wear and tear, identify problem parts or equipment, and prioritize large capital expenses. This can be a time-consuming task when performed manually. For computer-based systems, information can be stored in databases and retrieved quickly to expedite the process. There are several functions of these systems:

- Storing all relevant equipment nameplate data,
- Generating work orders,
- Tracking inventory and warehousing operations,
- Tracking maintenance history,
- Generating purchase orders,
- Keeping records of maintenance activity, and
- Reporting.

Computer-based budgeting and accounting functions are necessary to optimize budget control and analysis and to evaluate maintenance program performance. Analysis of flow data, rainfall, pump running times, and power consumption are all done on a routine basis to assist in data management.

3.0 CAPITAL PLANNING

A CIP is a multiyear planning instrument used to identify capital projects that are needed to increase the capacity of the system and to rehabilitate or replace infrastructure that is deteriorated or damaged from overuse or inadequate maintenance or repair. In addition, a CIP aims to coordinate the financing and timing of these improvements in a way that maximizes the return to the public.

Increasingly, professionals and policy makers are becoming aware that infrastructure exists as an interconnected system and that efficient functioning of the system is crucial to the environment, economy, and quality of life. The wastewater collection system, which includes pipelines, conduits, pumping stations, force mains, and other collection system-related facilities, is one of the largest infrastructural assets a community owns. Both public and private wastewater companies are being challenged to meet state and federal regulations, increased operating costs, and the expectations of customers for higher standards, while facing a deteriorating infrastructure and funding constraints. Collection systems degrade continuously, and a plan is necessary to effectively manage this degradation and prevent collection systems failures caused by pipe aging. Deferred maintenance will only increase costs for later generations and may cause catastrophic results.

The legal context under which municipal utility decisions have historically been assessed is that of utility law. The standard holds municipalities to a duty similar to the one imposed on a privately owned utility; that is, that service be supplied to all customers in similar positions to those already being served. Courts analyze municipal decisions relating to extensions of service on the basis of reasonableness and whether or not the extension discriminates against any individual or group of property owners.

Although utility law currently is the common law rule applied in most courts, legal emphasis is being placed on municipal planning for growth and regional coordination of planning. Several states have complex legislation that links capital improvement planning directly to comprehensive plans. Some courts have recognized the value that comprehensive planning considerations have to utility decisions. Whether or not a community is under the stress of rapid growth, the incentive to plan for future growth is present in nearly all of the states.

A CIP provides an excellent basis for making the link between growth management plans and utility policy. It allows a community to plan for future development by realistically assessing utility obligations, while formulating its policy to reflect the community's growth goals.

An integrated CIP should carefully plan for extension and expansion of collection systems. An increasing number of municipalities are requiring the costs of system growth to be placed on new users. As financing of new service and system rehabilitation becomes a budgetary burden on localities, the CIP can provide a sound basis for equitably and legally distributing the costs of system expansion.

In light of continuing changes in the legal importance of planning for growth, it is clear that state courts will have to address the conflict between utility law standards and the increasing need for municipal planning and growth management. When dealing with a challenged municipal utility policy, courts will have to determine a basis for that policy and may run up against conflicting objectives for collection system and water provisions. Objectives must balance growth management principles with what can be legally recognized as "utility-related" rationale. If such a policy is clearly outlined by a CIP that integrates utility cost and revenue projections with local growth management objectives and land use design, then a reasonable basis for judicial sanction will become clear.

Utility providers will not be able to operate in a vacuum. Good planning practices bring together affected stakeholders to determine how they want to prioritize their limited resources among level of service, the environment, and public health concerns. These concerns vary from community to community and from system to system. Planners and designers must extend their thinking beyond design tables and standardized approaches. *Guide for Municipal Wet Weather Strategies* (WEF, 2013) helps stakeholders understand how others have approached the planning process and evaluated the challenges that utility providers face. As service provision becomes more expensive and more important to metropolitan growth, an integration of values must form the basis for future utility extensions, rehabilitation, and financial policy. A CIP is based on the standard of service required by the customer. The CIP provides a means to protect, maintain, or improve the asset value of a collection system with planned repair, rehabilitation, or replacement on predicted deterioration of the system. In either a private or public utility, key information is needed to manage costs through asset-management planning (Kingdon, 1995). This information includes current conditions and performance of assets, operating costs, and financial position including revenues, balance sheet, and cash flow; required and anticipated future levels of service; and methods of measuring and monitoring performance of the system.

The primary goals of collection systems operation are to prevent public health hazards and to transport wastewater uninterrupted from source to treatment facility. The goal of a CIP is to predict the costs necessary to protect, maintain, or improve the value of this asset (the collection system) while providing the level of service desired. To meet these goals, the following factors must be addressed:

- Determining existing conditions,
- Setting future goals,
- Attaining future goals, and
- Tracking progress.

3.1 Master Planning

A master plan is a key component for developing a comprehensive CIP. Master plans are developed and updated to address system deficiencies and future growth. The basic goals of a master plan include

- Developing a long-term vision and strategy for management of the wastewater system.

- Identifying existing and future system deficiencies.
- Identifying and prioritizing system improvement projects to address current and future needs.

Master plans typically view the system through a predefined "build-out" year, often determined as part of a general plan or other community or utility master plan document. Most master plans first identify existing deficiencies in the system based on current system conditions and flows. Examples of existing conditions include capacity issues, severe I/I, overflows, and areas of known maintenance problems. Based on these existing systems analyses, projects are identified and added to the CIP. The second step is to reevaluate the system at or near build-out for potential zoning considerations of land redevelopment or repurposing. A plan would be to identify what components of the system need to be replaced or upsized for future flows. A hydraulic model as described in previous sections is a useful tool in this type of analysis. This analysis identifies the incremental improvements needed to address the future flows and growth to the system. This long-term approach provides a detailed roadmap for significant improvements needed for the system through build-out and to estimate the funds needed to implement them. Typically projects identified in a master plan become the "base list" of projects when developing a CIP. As described later in this chapter, it will be important to identify the priority and timing of projects as they are incorporated into the CIP.

If sufficient data is available, a master plan may include condition and other forms of asset management data. The inclusion of non-capacity-related project drivers to the standard master planning process may allow for the identification of efficiencies in the development of a comprehensive CIP.

3.2 Capital Improvement Plan Development

A typical city charter requires that the city manager prepare and submit a CIP for the 5-year period following the new budget year as part of the annual package. The CIP is a planning document and typically does not authorize or fund any projects.

Before the city manager brings forward a CIP, preliminary reviews typically take place to verify the accuracy, sustainability, and priority of projects selected. In most cases, this is done as a progression from the section (i.e., wastewater) to the department (i.e., utilities) to the governing board.

3.2.1 Ratings

To optimize decision-making and to help prioritize all potential solutions to a problem, it is important to incorporate the information that was collected using the methods described above and grouping them into three categories:

- Condition

- Performance
- Risk

This information can be converted into ratings to compare one scenario against another. Although cost is an important factor in identifying a solution, it is not the only factor that should be considered when developing a CIP. There are many ways to develop a scoring scenario, including the one described here. Each municipality will be different and should modify its approach to fit the agency's established business model.

The scoring procedure shown in Figures 4.2 and 4.3 is a method that can be used to look at various aspects of an asset and normalize them to make prioritization decisions. The example takes into consideration condition, performance, and risk, which are described in more detail below. By applying this type of methodology, every asset can be given a score (weighted appropriately based on each utilities' defined level of service). Once each asset has a score, GIS can be used to group the assets into projects resulting in a total score; Figure 4.4 includes this total project score. Having this score provides backup and justification for why one project is more important than another when presenting to management or the agency's governing body.

3.2.1.1 Condition

Condition scores are derived from the various tools and condition assessment methods described in previous sections. For example, if a pipe had sufficient capacity but most joints were offset or had excessive root intrusion, the condition score would reflect poorly. Examples of condition scoring components are

- Remaining service life,
- Condition of pipe (based on physical inspections),
- Number of repair service requests for particular assets (tracked in a CMMS), and
- Required cleaning or maintenance frequency.

An example of how these components could be combined to give an overall pipe rating for use in prioritization of CIP projects is shown in Figure 4.2.

3.2.1.2 Performance

Performance scores are derived from the various tools and performance assessment methods described in previous sections. An example of a poor

FIGURE 4.2 Capital Improvement Plan Prioritization Overall Scoring Methodology (WO = work order)

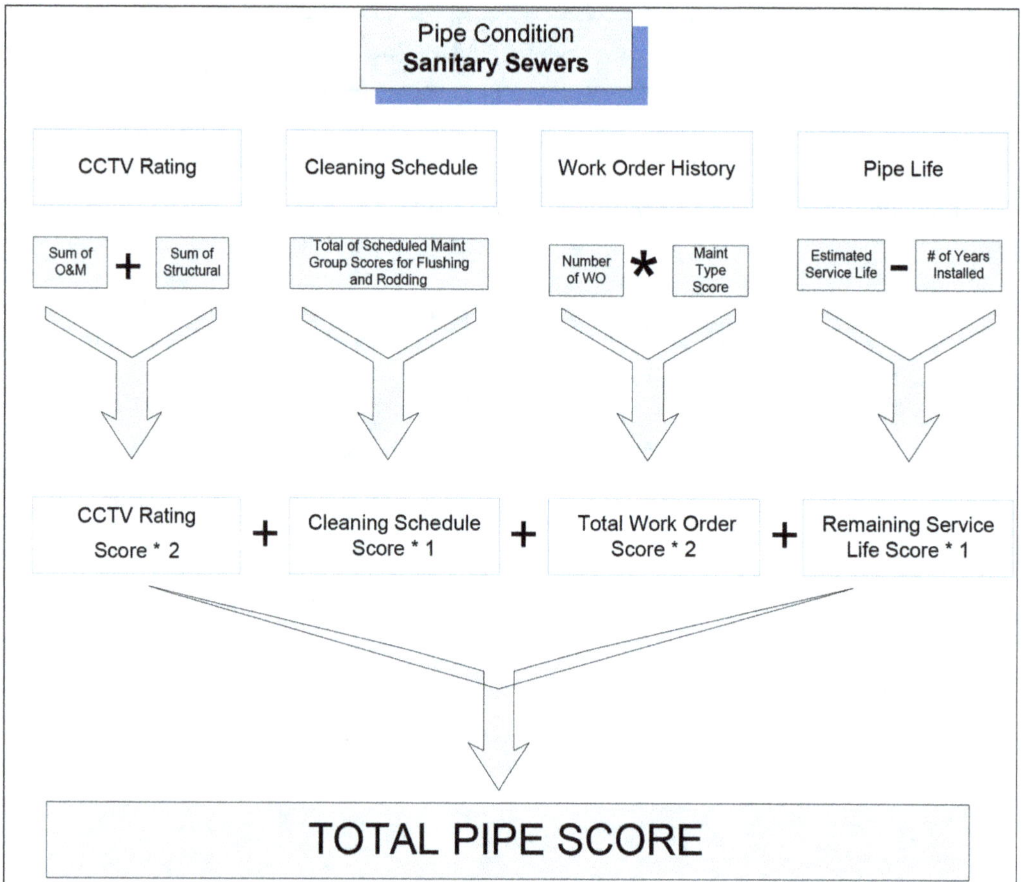

CCTV Pipe Rating	Score
< 5	1
5 - 15	2
15 - 25	3
25 - 40	4
>40	5
Remaining Service Life	**Score**
>40	1
25 - 40	2
15 - 25	3
10 - 15	4
<10	5

Cleaning Schedule	
Flushing	**Score**
1 Year	1
9 month	2
6 month	3
3 month	4
Rodding	**Score**
1 Year	3
6 month	4
3 month	5

WO Maint Type	Score
Unscheduled	2
Corrective	3
Emergency	3
Pipe Failures	**Score**
Deterioration	5
Construction	5
XXXX	5

FIGURE 4.3 Capital Improvement Plan Prioritization for Sanitary Sewer Pipe Condition (WO = work order)

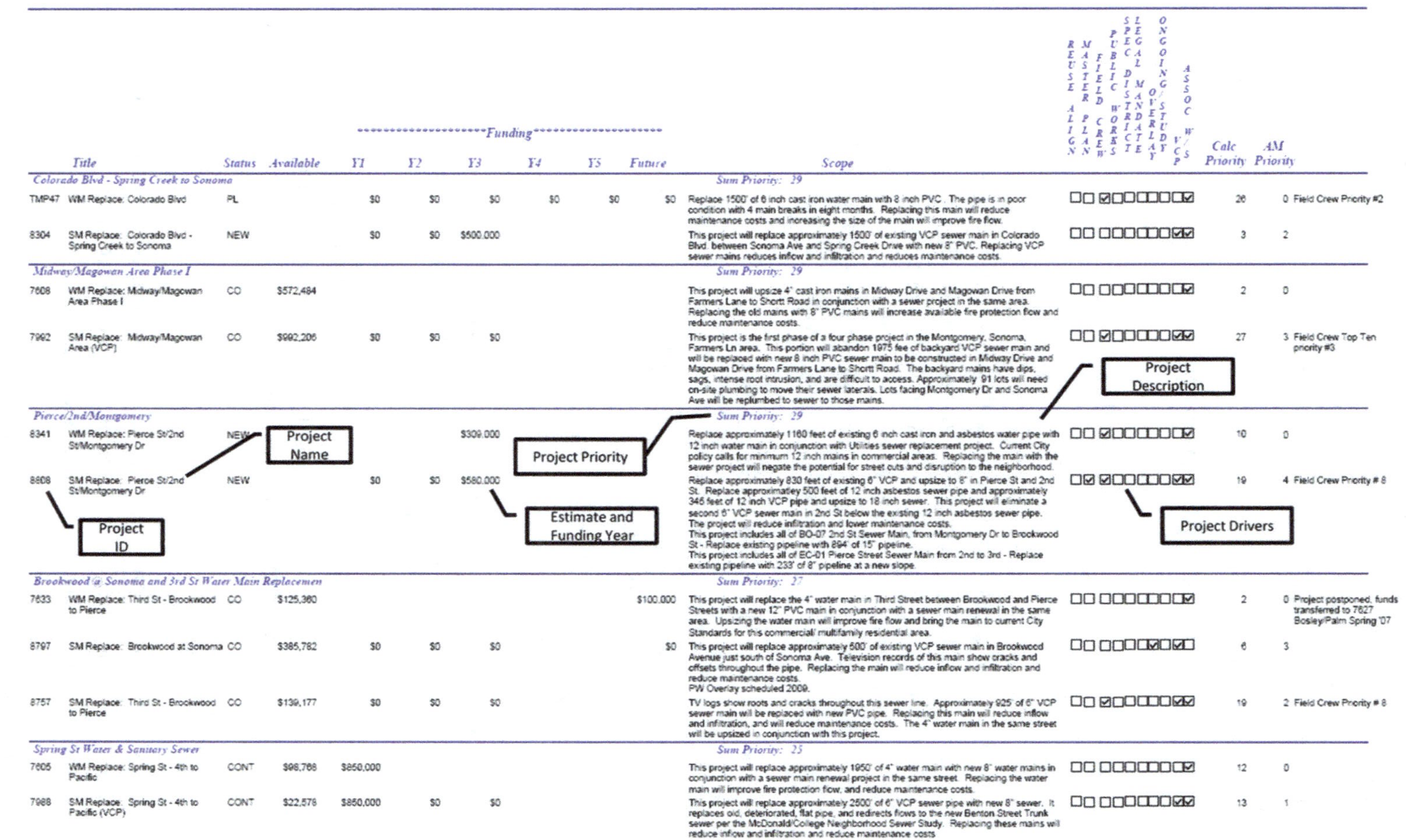

	Title	Status	Available	Funding Y1	Y2	Y3	Y4	Y5	Future	Scope	REUSE ALIGN	MASTER PLAN	FIELD CREW	PUBLIC WORKS	SPEC DISTRICT	LEGAL MANDATE	OVERLAY	ONGOING/STUDY	VCP	ASSOC W/S	Calc Priority	AM Priority	
Colorado Blvd - Spring Creek to Sonoma										*Sum Priority: 29*													
TMP47	WM Replace: Colorado Blvd	PL		$0	$0	$0	$0	$0	$0	Replace 1500' of 6 inch cast iron water main with 8 inch PVC. The pipe is in poor condition with 4 main breaks in eight months. Replacing this main will reduce maintenance costs and increasing the size of the main will improve fire flow.	☐	☐	☑	☐	☐	☐	☐	☐	☐	☑	26	0	Field Crew Priority #2
8304	SM Replace: Colorado Blvd - Spring Creek to Sonoma	NEW		$0	$0	$500,000				This project will replace approximately 1500' of existing VCP sewer main in Colorado Blvd. between Sonoma Ave and Spring Creek Drive with new 8" PVC. Replacing VCP sewer mains reduces inflow and infiltration and reduces maintenance costs.	☐	☐	☐	☐	☐	☐	☐	☐	☑	☑	3	2	
Midway/Magowan Area Phase I										*Sum Priority: 29*													
7608	WM Replace: Midway/Magowan Area Phase I	CO	$572,484							This project will upsize 4" cast iron mains in Midway Drive and Magowan Drive from Farmers Lane to Shortt Road in conjunction with a sewer project in the same area. Replacing the old mains with 8" PVC mains will increase available fire protection flow and reduce maintenance costs.	☐	☐	☐	☐	☐	☐	☐	☐	☐	☑	2	0	
7992	SM Replace: Midway/Magowan Area (VCP)	CO	$992,206	$0	$0	$0				This project is the first phase of a four phase project in the Montgomery, Sonoma, Farmers Ln area. This portion will abandon 1975 fee of backyard VCP sewer main and will be replaced with new 8 inch PVC sewer main to be constructed in Midway Drive and Magowan Drive from Farmers Lane to Shortt Road. The backyard mains have dips, sags, intense root intrusion, and are difficult to access. Approximately 91 lots will need on-site plumbing to move their sewer laterals. Lots facing Montgomery Dr and Sonoma Ave will be replumbed to sewer to those mains.	☐	☐	☑	☐	☐	☐	☐	☐	☑	☑	27	3	Field Crew Top Ten priority #3
Pierce/2nd/Montgomery										*Sum Priority: 29*													
8341	WM Replace: Pierce St/2nd St/Montgomery Dr	NEW				$309,000				Replace approximately 1160 feet of existing 6 inch cast iron and asbestos water pipe with 12 inch water main in conjunction with Utilities sewer replacement project. Current City policy calls for minimum 12 inch mains in commercial areas. Replacing the main with the sewer project will negate the potential for street cuts and disruption to the neighborhood.	☐	☐	☑	☐	☐	☐	☐	☐	☐	☑	10	0	
8808	SM Replace: Pierce St/2nd St/Montgomery Dr	NEW		$0	$0	$580,000				Replace approximately 830 feet of existing 6" VCP and upsize to 8" in Pierce St and 2nd St. Replace approximatley 500 feet of 12 inch asbestos sewer pipe and approximately 346 feet of 12 inch VCP pipe and upsize to 18 inch sewer. This project will eliminate a second 6" VCP sewer main in 2nd St below the existing 12 inch asbestos sewer pipe. The project will reduce infiltration and lower maintenance costs. This project includes all of BO-07 2nd St Sewer Main, from Montgomery Dr to Brookwood St - Replace existing pipeline with 894' of 15" pipeline. This project includes all of EC-01 Pierce Street Sewer Main from 2nd to 3rd - Replace existing pipeline with 233' of 8" pipeline at a new slope.	☐	☑	☑	☐	☐	☐	☐	☐	☑	☑	19	4	Field Crew Priority # 8
Brookwood @ Sonoma and 3rd St Water Main Replacemen										*Sum Priority: 27*													
7633	WM Replace: Third St - Brookwood to Pierce	CO	$125,360						$100,000	This project will replace the 4" water main in Third Street between Brookwood and Pierce Streets with a new 12" PVC main in conjunction with a sewer main renewal in the same area. Upsizing the water main will improve fire flow and bring the main to current City Standards for this commercial/ multifamily residential area.	☐	☐	☐	☐	☐	☐	☐	☐	☐	☑	2	0	Project postponed, funds transferred to 7627 Bosley/Palm Spring '07
8797	SM Replace: Brookwood at Sonoma	CO	$385,762	$0	$0	$0			$0	This project will replace approximately 600' of existing VCP sewer main in Brookwood Avenue just south of Sonoma Ave. Television records of this main show cracks and offsets throughout the pipe. Replacing the main will reduce inflow and infiltration and reduce maintenance costs. PW Overlay scheduled 2009.	☐	☐	☐	☐	☐	☐	☑	☐	☑	☐	6	3	
8757	SM Replace: Third St - Brookwood to Pierce	CO	$139,177	$0	$0	$0				TV logs show roots and cracks throughout this sewer line. Approximately 925' of 6" VCP sewer main will be replaced with new PVC pipe. Replacing this main will reduce inflow and infiltration, and will reduce maintenance costs. The 4" water main in the same street will be upsized in conjunction with this project.	☐	☐	☑	☐	☐	☐	☐	☐	☑	☑	19	2	Field Crew Priority # 8
Spring St Water & Sanitary Sewer										*Sum Priority: 25*													
7605	WM Replace: Spring St - 4th to Pacific	CONT	$98,768	$850,000						This project will replace approximately 1950' of 4" water main with new 8" water mains in conjunction with a sewer main renewal project in the same street. Replacing the water main will improve fire protection flow, and reduce maintenance costs.	☐	☐	☐	☐	☐	☐	☐	☐	☐	☑	12	0	
7988	SM Replace: Spring St - 4th to Pacific (VCP)	CONT	$22,576	$850,000	$0	$0				This project will replace approximately 2500' of 6" VCP sewer pipe with new 8" sewer. It replaces old, deteriorated, flat pipe, and redirects flows to the new Benton Street Trunk sewer per the McDonald/College Neighborhood Sewer Study. Replacing these mains will reduce inflow and infiltration and reduce maintenance costs	☐	☐	☐	☐	☐	☐	☐	☐	☑	☑	13	1	

FIGURE 4.4 Prioritized Total Project Score Example

performing pipe would be a pipe that was unable to satisfy the level of service expected or set by the agency. A pipe could be in pristine condition, but if it cannot handle the design flow, its performance score would be poor. Examples of condition scoring components are

- Hydraulic capacity,
- Design standards (slope, velocity),
- Flow characteristics, and
- Performance with build-out flows.

3.2.1.3 Risk

Identifying what risk is involved, will help identify what exposure the utility will have and help determine funding priorities. If a particular asset has a high risk to the agency or community if it were to fail, additional weight should be given. A pipe in poor condition that is low risk should be evaluated differently than a pipe in poor condition that is high risk for the agency. Examples of high business risk exposure would include

- Assets at the end of expected life cycle,
- Proximity to sensitive areas (highways, creeks),
- Absence of redundancy or backup for a specific asset that could result in high cost of failure,
- Poor reliability scores, and
- Imminent significant failure mode of capacity or level of service.

Risk also is defined as the relationship between the probability of failure and the consequence of failure. More details on how this process is used in project validity can be found in WERF's strategic asset-management tool Sustainable Infrastructure Management Program Learning Environment.

By bringing together these various scoring elements, it is possible to develop a total score using weighting factors to satisfy particular utility goals and concerns. An example of how this process works is shown in Figure 4.2.

3.2.2 Validation

Once the projects and assets have been identified as candidates for rehabilitation, renewal, or replacement, it is important to validate the project. Questions to ask about a project include

- Should the project be eliminated?

- Can the project be deferred?
- Can maintenance activities be adjusted or modified?
- Can the operation of the asset be changed?
- Is there a non-asset solution?

Answering these questions will help the utility determine if the specific project is the right solution, at the right cost, scheduled at the right time, and for the right reasons. A lot of work has been done to develop methods for assessing the validity of a project. One way is to develop a business-case evaluation that looks at all aspects of the project to ensure the right solution is being proposed. In the past, utilities looked only at the financial effects of a project. This is changing as more utilities are using the triple-bottom-line method, which takes into account the financial, social, and environmental effects of the project.

3.2.3 Replacement/Rehabilitation Goals

Replacement goals are one way a municipality can identify the investment required to reach goals identified in the CIP or by governing boards. A replacement program helps facilitate the long-term budgetary planning and helps the municipality plan a gradual replacement of the aging collection system infrastructure in a way that is palatable to management, governing boards, and the public.

A basic method for determining a replacement goal is to identify the pipe age and serviceability as described previously. Where pipe material or data is unknown, assumptions will need to be made based on location and interviews with field operations staff. In this basic method, replacement because of age and serviceability is assumed to be required at the end of a pipeline's average service life based upon industry standards and past experience by operations and engineering staff for a particular municipality. Using this method, it is possible to estimate the quantity of pipe that will exceed service life by a chosen date. This can then be converted into a replacement rate. Typically, municipalities will need to have a more aggressive replacement rate initially to bring the system up to a point where a more stable and palatable rate (i.e., 1%) can be applied in future years to accommodate pipe with a longer life cycle (i.e., more than 100 years).

3.2.4 Industry Benchmarks

Benchmarking is a process that management can use to gauge performance of a collection system and to define the goals and priorities of a CIP. Many large corporations have used benchmarking, and the fundamental principles affecting productivity, profitability, or organization are similarly valid for wastewater collection systems. The benchmarking process is straightforward, as shown in the following six steps:

1. Identify benchmarking functions,
2. Identify performance criteria,
3. Identify best-in-class utilities and measure best-in-class performance,
4. Measure system performance,
5. Close the gap between measured system and best-in-class performance, and
6. Implement changes.

When identifying benchmarks, those functions that can be measured easily and that allow managers to exert influence through changes in policies and procedures should be targeted. Identification of performance criteria should include consideration of both financial measures (such as return on assets) and strategic measures (such as customer-perceived quality). When performance is measured, the organizations being compared should have the same variables. These performance indicators are sometimes needed for wastewater collection systems to justify budgets. Any improvement plans to close the gap between performance of comparable utilities must be weighed against the cost and investment required to achieve the desired results. Finally, implementation of any changes should consider the time needed to achieve results and include a system to monitor and reassess results.

Benchmarking is defined as a method of comparing a utility or operation with a selected group of similar operations. The AWWA has developed a program for service providers to compare relative performance through benchmarking. Details of this program can be found at https://www.awwa.org/Resources-Tools/Programs/Benchmarking. The indicators developed were designed to help utilities improve their operational efficiency and effectiveness by assessing their own utility against a standardized benchmark. Another goal of benchmarking is to provide an expanding inventory and condition assessment over time. The results can then be used for prioritization and justification of a CIP. However, participation in the survey is voluntary and therefore is not necessarily representative of the entire industry. Anecdotally, over the years it appears that more recently only good performers are responding; therefore the mean or average has been elevated.

3.2.5 Prioritization

This chapter has described various methods and tools used to determine the condition of a collection system. It is important that a manager take all this information and prioritize it in a way that will produce a CIP to

- Improve water quality by reducing overflows,
- Reduce costs by minimizing required emergency responses and repairs,
- Reduce costs by improving management and extending system life,
- Develop sufficient system capacity for existing and future growth,
- Substantially reduce the amount of I/I entering the system, and
- Identify the most cost-effective method for addressing system needs.

To help the manager prioritize the vast amount of information identified using the tools and methods described in this chapter, a decision matrix should be developed. A decision matrix identifies system metrics that are important to the utility and assigns a weighting factor according to their respective value identified by management and staff. Metrics that should be included in developing a decision matrix (in no particular order) include

- Capacity deficiency;
- Inflow and infiltration;
- Known O&M problems;
- Design flow;
- Project cost;
- Project coordination;
- Service life;
- Required by regulation;
- Identified in master plan; and
- Socioeconomic impacts, environmental improvement, and others identified by stakeholders.

3.2.6 Final Product

After performing a system analysis using the tools described above and identifying a plan incorporating master planning, prioritization, and agency goals, it is time to develop a plan to communicate the CIP to staff, governing body, and the public. A well-organized, complete CIP will include the following elements:

- Project name and identification,
- Project type or category,
- Budget estimate,
- Funding year,
- Priority rank (based on risk, condition, and performance), and
- Description of project.

Although every CIP will have unique layouts and varying information, the elements above represent a standard of best practices for a typical CIP. An example of a CIP with these elements is shown in Figure 4.4. The CIP is a planning document and does not authorize or fund any projects.

A significant component of this plan is the budget estimate and funding year. This information is used to help determine the most appropriate funding strategy to accomplish the plan. For a more detailed description on this process, refer to Chapter 8, Budgeting and Financial Planning.

4.0 REFERENCES

American Water Works Association Research Foundation & Water Environment Research Foundation. (2007). *Condition assessment strategies and protocols for water and wastewater utility assets.* Water Environment Research Foundation; AWWA Research Foundation.

Kingdon, W. D. (1995). Water/Engineering and Management. Water Research Center.

Water Environment Federation. (2013). *Guide for municipal wet weather strategies.*

Water Environment Federation. (2017). *Sanitary sewer systems: Rainfall derived infiltration and inflow (RDII) modeling.* https://www.wef.org/globalassets/assets-wef/direct-download-library/public/03---resources/wsec-2017-fs-001-rdii-modeling-fact-sheet---final.pdf

Water Environment Federation & American Society of Civil Engineers. (2008). *Existing sewer evaluation and rehabilitation* (3rd ed., WEF Manual of Practice No. FD6/ASCE Manuals and Reports on Engineering Practice No. 62). McGraw-Hill.

5.0 SUGGESTED READINGS

American Water Works Association Benchmarking Program. (n.d.). https://www.awwa.org/Resources-Tools/Programs/Benchmarking

Harlow, K. (2007, November/December). Beyond risk: A community return on investment approach to pipe rehabilitation. *Underground Infrastructure Management*, 38–40.

Kirby, R., Cummings, R., & DiTullio, B. (2007, September/October). Asset management in practice: Lessons learned from Hillsborough County. *Underground Infrastructure Management*, 25–29.

Lafferty, A. K., & Lauer, W. C. (2005). *Benchmarking performance indicators for water and wastewater utilities: Survey data and analyses report.* American Water Works Association.

Linschoten, G. (2007, October 13–17). Lifecycle analysis and capital planning using asset condition and deterioration. In *Proceedings of the 80th Annual Water Environment Federation Technical Exposition and Conference* [CD-ROM], San Diego, CA.

National Association of Sewer Service Companies. (n.d.). www.nassco.org

Water Environment Federation. (2006). *Guide to managing peak wet weather flows in municipal wastewater collection and treatment systems.*

Water Research Foundation. (n.d.). *Sustainable infrastructure management program learning environment.* https://www.waterrf.org/resource/sustainable-infrastructure-management-program-learning-environment-simple-foundation

5

System Design Considerations

1.0 INTRODUCTION

This chapter considers design factors and how they affect operations and maintenance (O&M) for wastewater systems. Many of the topics discussed in this chapter also pertain to storm and combined sewer systems. However, the inclusion of specific design issues for storm and combined sewers is beyond the intent of this manual.

Criteria used in a particular wastewater design typically are established in the preliminary stages of the project. It is advisable to seek input from internal design staff, maintenance managers, superintendents, operators, geotechnical engineers, and construction personnel during the development of design and routing criteria. People in these positions know where field problems are likely to be encountered; thus, they can offer valuable information in the early stages of the project.

Key design considerations that affect O&M include

- Line locations, grades, depth, and material;
- Line clearance under waterway flow lines and clearance away from waterway banks;
- Route geography;
- Manhole spacing and size;
- Manhole channels or inverts;
- Influences of lift stations;
- Odor control provisions;
- System instrumentation and controls;
- Induced turbulence (such as drop manholes);
- Wastewater and manhole accessibility;
- Allowance for inflow and infiltration (I/I) capacity;
- Knowledge of system flow characteristics;
- Initial wastewater flowrates compared to design flowrate;
- Degree of construction observation; and
- Proficiency of the wastewater maintenance department.

Neither routine maintenance nor emergency service can be successful if maintenance personnel do not have appropriate access. One way to encourage the designer to account for operational considerations is to require the maintenance manager's approval on conceptual and final design plans.

Ideally, collection systems should be designed to minimize total life-cycle costs; that is, the initial cost plus the present value of future O&M costs. This implies that a well-designed system can be easily and efficiently operated and maintained. However, some well-designed systems are not constructed as designed because of poor construction techniques, field changes, the absence of construction observation, and inadequate testing. Regardless of the reason, most collection systems will experience some deterioration and require general maintenance.

1.1 Regulatory and Environmental Requirements

1.1.1 National Pollutant Discharge Elimination System Permits

National Pollutant Discharge Elimination System (NPDES) permit language contains standard references to collection systems in four permit conditions: non-compliance reporting, proper operations and maintenance, duty to mitigate discharges, and prohibition and defense provisions. Designers should review these requirements with the collection system owner and consider whether state or federal clarification of the permit requirements is necessary. For instance, are the non-compliance reporting requirements clear? Are there exceptional circumstances in which prohibition and defense provisions need to be expanded or be more specific?

Entities that operate and maintain a collection system, whether or not they have an NPDES permit, are responsible for any wastewater discharges to waters of the United States. Municipalities or permit holders should verify that local ordinances are in place that clearly establish the responsibilities and controls expected from flow-contributing private laterals and private satellite collection systems that are served and do not have their own NPDES permit.

1.1.1.1 Sanitary Sewer Overflows and Combined Sewer Overflows

The U.S. Environmental Protection Agency (U.S. EPA) and local state organizations have developed proactive measures to reduce sanitary sewer overflows (SSOs) and combined sewer overflows (CSOs) in recent years. These measures have led to the development of programs and enforcement actions to not only reduce overflows, but to encourage wastewater system owners to actively prevent SSOs/CSOs through financial penalty.

1.1.1.2 Capacity, Management, Operations, and Maintenance

To help eliminate and control SSOs and CSOs, U.S. EPA developed a capacity, management, operations, and maintenance (CMOM) program designed to ensure collection systems have effective programs in place. Moreover, it could provide a means of reporting SSOs from collection systems. Although U.S. EPA has not officially taken on the role of regulating CMOM for collection systems at the national level, regional EPA offices, state, and local agencies are serving in this capacity.

Most states in the U.S. have implemented a CMOM program, requiring all collection systems in the state to prepare sewer system management plans (SSMP) and report SSOs in a statewide reporting system. These plans in general require the collection system operator to provide the following 10 elements, which are all elements of the U.S. EPA CMOM guidance:

- Goals;
- Organization;
- Overflow emergency response plan;
- Fats, oils, and grease control program;
- Legal authority;

- Measures and activities;
- Design and construction standards;
- Capacity management;
- Monitoring, measurement, and program modifications; and
- Collection system management plan audits.

If the agency feels any element is not applicable, it merely needs to address that in its SSMP per local guidance. The program is a monitoring and reporting tool that will not only track SSOs and CSOs, but ensure that collections systems are operated and maintained properly. For further information, please visit https://www.epa.gov/npdes.

1.1.2 System Conveyance Design Condition

Collection systems, or major hydrological sections of the system, should have a design flowrate that meets or exceeds state or federal regulatory agency requirements. Many entities develop this design rate through application of a design storm as a level of service. These designs storms are typically related to a storm event that might return in 2 (50% chance in a given year), 5 (20% chance in a given year), or 10 (10% chance in a given year) years on average. Design rates should include the planning period's wastewater flow projections and an I/I allowance that reflects existing or targeted seasonal and rain event flow variances. Establishing an overall conveyance design flowrate will provide the basis for new construction and rehabilitation design criteria and help avoid hydraulic bottlenecks or constrictions in the system that can lead to SSOs.

1.1.2.1 Risk-Based Design Options

Recently there has been a trend to look at collection system design flowrates from a risk-based rather than storm event–based perspective. This approach allows a community to decide whether it is prudent to spend the dollars for a system able to handle the design storm event now or to install a system that would handle the majority of flows and be less expensive. This risk analysis compares the costs and risks associated with controlling or not controlling flows from a large future storm event for different scenarios. Further discussion may be found in the Water Environment Federation's *Wastewater System Capacity Sizing Using a Risk Management Approach* (2007).

1.1.2.2 Wet-Weather Facilities

Wet-weather facilities are designed and constructed to manage or control flows associated with wet-weather events. These facilities help protect the environment and public health. They may be designed to discharge under certain conditions upstream of a wastewater treatment facililty. Any facility designed with an anticipated discharge must be permitted; however, wet-weather facilities may be designed for off-line or in-line storage with no direct discharge to a waterbody.

Wet-weather facilities may include one or a combination of processes, including screening, primary sedimentation, in-line or off-line storage basins, chlorination, or dechlorination (site-specific stream conditions may not require storage). For example, the facility will be designed to function according to site-specific, wet-weather events whereby the collection system wet-weather peak flows are diverted to wet-weather facilities for storage. If the diverted flowrate or volume is greater than what the facility can handle, excessive flow is discharged to a receiving stream after having gone through as much of the permitted process layout as possible. When wet-weather flows diminish, the stored wastewater is then released back to the collection system for treatment.

Wet-weather facilities can help prevent problems associated with basement backups and reduce, through their containment functions, the frequency of SSOs. Wet-weather facilities must undergo a comprehensive evaluation and justification process, but can be the most reasonable, cost-effective alternative to larger conveyance systems that are unable to achieve adequate cleansing velocities during normal flow conditions.

1.1.3 Design Approvals

Conveyance system design approval procedures vary between states and localities. The designer should first discuss the design standards to be used with the owner of the collection system. The designer should contact state, county, or other agencies to learn of any mandatory design standards or criteria that need to be addressed to receive regulatory approval. Owner agency design standards can only be equal to or more stringent than the state or county standards.

1.1.4 Environmental Issues and Compliance

New construction and rehabilitation in collections systems typically can affect several different aspects of the environment:

- Endangered species;
- Air quality;
- Traffic;
- Protected lands;

- Archeological sites;
- Waterways, vegetation, and paved areas; and
- Water quality.

Environmental issues should be considered during design so that mitigation is understood and incorporated in the contract documents.

1.2 Permitting

Permit requirements also affect design. The designer should fully understand applicable permits that will be required and the stipulations of such permits before finalizing their design. Permits can range from local encroachment to U.S. Army Corps of Engineers 404 permits. It is a good idea to discuss permitting with the owner's engineering staff because they may have a better understanding of requirements.

2.0 DESIGN GUIDELINES

2.1 Design Process Checklist

To aid the designer, a checklist should be used to ensure the necessary steps are followed:

- Project manager and team are assigned;
- Project intent clarified (new development, replacement, rehabilitation, etc.);
- Project schedule and budget established;
- Route selected;
- Alignment study completed for pipeline in selected route;
- Public, traffic, and environmental effects understood;
- Permit requirements considered;
- Survey conducted;
- Utility conflicts identified and potholing complete;
- Geotechnical investigations completed;
- Hydraulics and pipe size determined;
- Pipe material selected;
- Manhole type and spacing requirements established;
- Pipe installation configuration and materials completed;

- Laterals and connections defined;
- Traffic control planned;
- Corrosion and odor control in place;
- Right-of-way needs and staging areas addressed;
- Special design features like tunnels, structures, and appurtenances incorporated;
- Input from O&M personnel, especially future access needs considered;
- Basis of design report established;
- Design submittals at conceptual, preliminary, and final stages completed;
- Value engineering and constructability after conceptual stage if required;
- Appropriate quality control and quality assurance performed;
- Cost estimated;
- Specifications outlined;
- Bidding requirements clarified; and
- Construction supported.

2.2 Project Management

To have a successful project that meets the design intent, the project manager will have to help control schedule and budget and ensure that the final product satisfies the original requirements. This includes assigning proper personnel to assist with the aspects of the project and that tasks are accomplished at predetermined milestones. The project manager will also need to ensure that the quality of the documents is maintained and obtain required approvals. The manager will lead the remainder of the team through a planned process to complete the project. Tools used by the project manager might include scheduling software, budget tracking using a spreadsheet or database, presentation software, and communication (email, etc.) software. The manager will ensure that each component of the project is completed at its prescribed time and of the right quality.

2.3 Using Consultants

Consultants have been used successfully to handle project management as well as other aspects of design. Typically there is a qualification phase, a proposal phase, and a selection phase that may or may not include interviews with the team. Perhaps the most important aspect in the selection process is to select a firm and team that has completed recent and similar projects.

2.4 Routing Considerations

2.4.1 Assessing Aboveground and Belowground Interference

Routing considerations will be different for gravity versus pressure sewers, but both should emphasize O&M accessibility in their design criteria. Many factors affect the final selection of a route. Important considerations for aboveground systems include

- Paved versus earthen surfaces;
- Surface water flooding potential;
- Proximity to residential and commercial buildings;
- Construction problems such as limited equipment movement or stability risks of adjacent structure;
- Traffic or recreational disruption;
- Biological, aquatic, or historical disruption; and
- Number of land parcels traversed and easement requirements.

Important considerations for belowground systems include

- Geology and soil conditions;
- Groundwater, river, or tidal influences;
- Other utility conflicts or influences (such as cathodic protection systems); and
- Depth of trench and safety requirements.

2.4.1.1 Surveys

It is important to ensure quality field data in conveyance design. It is imperative to obtain the best possible description of existing conditions, including existing utilities and surface features, as changes during construction may result in significant cost increases. Carefully selecting the surveyor based on previous experience is just as important as selecting the designer. Ideally, they have worked together in the past. Location of utilities aboveground may not be adequate and potholing may be necessary to identify vertical conflicts with the new construction..

2.4.1.2 Potholing

Potholing refers to locating underground utilities by vacuum extraction or other means to determine depth and horizontal alignment. It is critical in cases in which there are existing pressure pipelines such as water, gas, or explosive substances. However, fiber optic and communication lines also are critical; damage can affect an entire community. Adequate funds should be set aside to do a good job and locate as many utilities as needed. Additional funds spent in the design phase on potholing often reduces the overall cost of construction.

2.4.2 Geotechnical

Geotechnical surveys can reveal important information to the designer that will not only affect design and construction costs, but also will influence the life of the conveyance system. Deterioration of sewer and manhole materials can be accelerated if there are not adequate provisions for protection from corrosive soils. Surveys can reveal the extent of rock in an area and influence the feasibility of whether to install a gravity or pressure system. Geotechnical surveys also can reveal soil expansion and stability conditions so that the designer can select materials and systems that will maintain water tightness while allowing for some lateral or longitudinal movement within the pipe and bedding zone. Typically, geotechnical data should be obtained at intervals of 300 m (1000 ft) or more often along the collection system alignment, depending on the anticipated changes in soil characteristics and makeup. For tunneled crossings, geotechnical data should be obtained at the proposed beginning, middle, and end locations. This information can help determine the type of tunnel to build and equipment needed to construct it. In an open-trench construction, the geotechnical data can provide recommendations for sheeting and shoring requirements.

2.4.3 Traffic Control

Depending on the final location of the proposed collection system, it may be prudent to provide guidance to the contractor for minimizing traffic disruption. Flagging may not be enough protection. Managers should consult with the local agency having jurisdiction of the roadway to determine the standards that they expect to be followed for traffic control during construction.

2.4.4 Permanent Access

Many I/I and maintenance-related problems occur because a collection system is inaccessible and cannot be routinely evaluated. Permanent access roads or gate assemblies on unimproved land should be considered a part of new wastewater construction projects where possible. The cost for constructing these access roads is not likely to decrease in the future, and most utility contractors will have suitable materials and equipment on hand to perform this task. However, construction of an access road across a wetland area should be evaluated carefully. The decision of whether to provide permanent access must be coordinated with permitting and other interested groups because wetlands will vary in physical character. Some wetland areas may require too much time, effort, and cost to enable access. The manager should look for environmentally sensitive, less-expensive, and less-controversial alternatives.

The designer should work closely with the collection system owner and legal staff to discuss the required temporary and permanent easement widths. If permanent widths are too narrow for the collection system depth, repair costs are likely to be significantly greater in the future because of extra trench shoring and accessibility restrictions. This is particularly true as the area becomes more developed and encroachments into the easement occur.

Depending on the diameter of the proposed pipeline, the choice of initial routing location should minimize the amount of land parcels that are bisected. Designers and collection system owners tend to underestimate the time needed to obtain easements. It is advisable to have all easements obtained (or at least critical easements) before awarding the contract or beginning construction. A process and schedule for obtaining easements is crucial to the overall project.

2.4.5 Public Effects

Because the public is affected by these projects, their needs must be taken into account. Otherwise politics will drive the solution rather than engineering. When a project will affect the public—such as because of changes in traffic flow and patterns—it is best to hold public meetings and seek input. It is also useful and beneficial to communicate with local law enforcement and emergency response agencies when construction requires changes in normal traffic patterns.

2.4.6 Rights-of-Way and Easements

The designer needs to ensure proper access for construction and future maintenance personnel by developing cross-sections of pipeline alignment and identifying several key items:

- How the trench will be excavated with backhoe clearances.
- How spoil materials from the trench will be hauled away.
- How materials will be returned for bedding and backfill.
- Onsite storage and staging needs.

- Pipe delivery method and installation into the trench.
- Adequate clearances to maintain proper traffic control.
- Future maintenance traffic needs and access.
- Need for security.

Permanent easements must provide adequate room for future excavation in case of pipe replacement or rehabilitation. Vehicles and equipment will need to be able to access manholes easily for inspection and cleaning. Typically easements are centered on the installed pipeline; accessibility, however, should dictate location. It is not unusual to need approximately 9 to 14 m (30 to 45 ft) for pipeline access.

Temporary easements range in size depending on the project and can be 15 m (50 ft) wider or more on one side or both sides of permanent easements. For a 183-cm (72-in.) pipe, 37 m (120 ft) total easement width would not be unreasonable. It is also important to recognize that temporary easements may be needed to provide staging areas for equipment, materials, and supply storage. Depth of cover for the proposed pipeline is the largest factor to consider when evaluating temporary construction easement requirements.

Easements are property rights granted from one owner to another. If the pipeline will traverse a public right-of-way or street, easements are not required. Private property raises other considerations. If the owner will not grant the easement, consideration must be given to other recourse such as eminent domain (condemnation) or developing alternate routes.

2.5 Hydraulic Analysis Using Modeling

The designer will need to work with the system owner and their master plan to determine how the hydraulics should be handled. A discussion of the use of hydraulic computer models for system design and capacity analysis is provided in Chapter 4.

2.6 Lift Stations

2.6.1 Package Stations

Collection systems cannot always be designed as a complete gravity system all the way to the water resource recovery facility. In some smaller systems, the need for lift stations can be handled by purchasing package lift stations. These can range from submersible systems that can be installed in standard concrete manholes to more elaborate self-contained systems with full-service electrical gear. The wet well (location where wastewater enters and is stored) typically is the only major structural component. There are, however, package stations with both wet and dry wells (dry wells are used to keep pumps dry and out of the wastewater environment). Regardless of station type, it is good practice to provide redundancy and a backup pump in case of duty pump failure. Equipment manufacturers can provide design references for this type of lift station.

2.6.2 Other Types of Stations

Although the full design features for lift stations is beyond the scope of this manual, larger lift stations should contain several basic components:

- Wet well
  - With submersible pumps only (screening equipment not provided) and
  - With dry pit pumps (screens will be required ahead of wet well entrance)
- Drywell (if not submersible pumps)
- Screening or grinding equipment ahead of wet well entrance
- Pumps (redundancy should be provided by having at least one duty and one standby pump for the largest size pump provided)
  - Submersible in wet well and
  - Dry pit centrifugal or other
- Pump controls
- Electrical supply
- Supervisory control and data acquisition or other operating equipment

Lift station design is covered in more detail in *Pumping Station Design* (Jones et al., 2008).

2.7 Pipelines and Joint Materials

2.7.1 Gravity Pipelines

2.7.1.1 Sizing Considerations

Gravity pipeline sizing should be determined from applying design criteria to system conveyance-planning flow projections. In addition, there are many reference texts and manuals that provide concepts for sizing a gravity sewer, such as *Gravity Sanitary Sewer Design and Construction* (American Society of Civil Engineers [ASCE] & Water Environment Federation [WEF], 2007). The most common method of gravity pipeline design is the use of the Manning equation in which capacity or flowrate is based on cross-sectional area, hydraulic radius, slope, and roughness coefficient known as the "n" factor.

Typically Manning's "n" will vary, but 0.013 is used widely. However, selection of the pipe diameter is increasingly being influenced by O&M considerations, such as scouring velocities or providing wet-weather, in-line storage.

Some communities have adopted sewer criteria that establish a minimum velocity of 0.75 to 0.90 m/s (2.5 to 3.0 ft/sec) rather than 0.61 m/s (2 ft/sec) at half- or full-pipe flow conditions. Other communities have oversized their interceptors to provide in-line storage during wet weather and incorporated O&M activities to induce cleansing velocities. Camp's equation, as cited in the previous reference, can also be used to determine the cleansing velocity in collection systems:

$$V = [(8B/f)g(s - 1)D_g]^{1/2}$$

Where

V = velocity (m/s, ft/s),
B = constant ranging from 0.4 to 0.8 (dimensionless),
f = Darcy–Weisbach friction factor (dimensionless),
g = gravitational acceleration (9.8 m/s^2, 32.2 ft/s^2),
s = specific gravity (dimensionless), and
D_g = diameter of particle (m, ft).

or

$$V = (1.486/n)\ R^{(1/6)}\ [B(s - 1)D_g]^{1/2}$$

Where

V = velocity (m/s, ft/s),
n = Manning coefficient of roughness,
R = hydraulic radius (m, ft),
B = constant ranging from 0.4 to 0.8 (dimensionless),
s = specific gravity (dimensionless), and
D_g = diameter of particle (m, ft).

Increasingly available and affordable hydraulic modeling programs are aiding in the decision-making process regarding pipe and joint sizing.

2.7.1.2 Materials

Proper collection system design includes careful evaluation of construction materials. Future maintenance and rehabilitation costs can be reduced by addressing factors that affect the life expectancy and use of the collection system. Wastewater characteristics, site conditions, and construction methods affect material selection. Factors to consider when evaluating a gravity piping system include

- Life expectancy and performance history;
- Availability of sizes required;
- Backfill and bedding material requirements;
- Availability and ease of installation of fittings or connections;
- Type of joint (fluid and pressure tightness and ease of assembly);
- Resistance to scour or abrasion;
- Resistance to acids, alkalis, gases, solvents, and other industrial wastes;
- Ease of handling and installation;
- Physical strength characteristics;
- Flow characteristics (friction coefficient);
- Cost of materials, handling, and installation; and
- Manufacturer testing and quality control program.

No single material is optimal for all conditions; selection typically is made based on the particular application under consideration. Different grades of materials may be selected for portions of a single project, but this practice typically is kept to a minimum unless a significant cost savings can be attained. The differential in the costs of various types of pipe tends to be a small part of the total project cost but may be worth considering as a more expensive pipe and manhole material may reduce future O&M costs. *Gravity Sanitary Sewer Design and Construction* (ASCE & WEF, 2007) describes the more typically used materials and their advantages and disadvantages.

The basic pipe materials commercially available for gravity sewer construction and given in the references are

- Ductile iron,
- Reinforced concrete,
- Polyvinyl chloride (PVC),
- High-density polyethylene (HDPE),
- Fiberglass, and
- Vitrified clay.

Ductile iron and concrete pipes typically have interior linings or exterior coatings to extend their life expectancy because of the aggressive and corrosive sewer environments to which they are subjected. Ductile iron can also include a polyethylene bag or wrap to prevent external corrosion from aggressive soils.

2.7.1.3 Backfill and Bedding

For the purposes of this discussion, backfill will include the zone of material around the pipe in a trench installation. The area immediately above, below, and around the pipe will be called the pipe bedding. In practice, initial, base bedding is placed, the pipe is installed on that surface, and then final bedding is extended above the pipe to adequately protect it. Both the pipe backfill zone and the bedding area serve an important function in the pipe's ability to support the resultant load forces. The materials that can be used in the pipe zone and bedding area vary as widely as the soils in which they are buried. Therefore, it is important for the designer to coordinate the pipe's inherent strength characteristics with the type of backfill and bedding material that will be used during construction.

Attention to backfill and bedding requirements during design and construction can eliminate many typical O&M problems. Improper pipe backfill and bedding design or a lack of uniform bedding during construction can lead to pipe structural failures; differential joint settlement; pipeline sags; and admission of groundwater, soil, and roots. In addition to the pipe manufacturer's design guidance literature, there are several good references the designer can review for a more detailed discussion of backfill and bedding design practices, such as *Gravity Sanitary Sewer Design and Construction* (ASCE & WEF, 2007).

Designers should consider whether to include an allowance for future earthen loads when selecting a pipe and backfill combination. For instance, pipelines routed along undeveloped urban streams or ditches may experience in their design life several meters (feet) of fill material greater than the original earthen depth. This often happens when a developer wants to raise the foundation elevation of proposed new buildings above the flood elevation.

Deviation from designed trench or embankment widths during construction may result in an overload to the pipe. This is particularly true when a wastewater pipe approaches and leaves a manhole. Designers should be aware that if a limited trench width is required for the backfill material specified, and if it is less than the width that is required to place a manhole in the trench, the potential exists for an overloaded pipe. Also, inadequate backfill compaction could cause differential settlement, overdeflection, or shearing, which can lead to pipe failure. An evaluation of the probability of differential pipe settlement versus structure settlement is vital for a successful design.

Granular fines can be transported out of a compacted backfill if the terrain grades are steep. The designer should consider installing a solid encasement around the pipe or a water barrier such as clay, bentonite, or other impervious material in the trench to impede subterranean flow along the pipe. Fines can migrate from the native soil in the trench walls into the granular backfill as well. This can cause a loss of lateral support and additional load-bearing forces on the pipe. Flexible and semirigid pipe materials will be affected by the loss of lateral backfill support. If there is material lost within the pipeline's haunch, the pipeline will experience loads greater than what the designer had anticipated. One method to reduce fines migration is to line or wrap the trench bedding and backfill zones with a geotextile material (sometimes called a "burrito wrap"). This method is also helpful when native soils in the trench are soft organic clays or peat that might allow the granular material to move laterally when placed under load.

The O&M problems and pipe failures previously mentioned also can occur when the pipe diameter becomes sufficiently large that the installer has difficulty placing and compacting the backfill under the haunches of the pipe as required. This problem is more common for pipe diameters greater than 60 cm (24 in.). A flowable fill (grout) or controlled density fill is an alternative; however, other issues such as the potential to float the pipe and chemical attack to metal pipes and appurtenances become considerations that also must be addressed.

Insufficient cover over a pipe can cause pipeline problems similar to those previously discussed. Operations and maintenance problems may develop faster with shallow cover because the failed or defective pipe is more susceptible to roots- or rain-induced I/I. In colder regions of the country, pipes should be installed below the frost line.

2.7.1.4 Grades

The composition of domestic wastewater has changed as the public's needs and practices have changed. Many convenience wastes, such as non-returnable or disposable materials, find their way into collection systems. Proper pipe grades help keep these wastes from settling.

Too often, collection system designs include long lengths of pipeline installed at minimum grades, typically accepted as grades that produce 0.61-m/s (2-ft/sec) flows at half- and full-pipe diameters. A somewhat steeper grade that produces velocities of 0.91 m/s (3 ft/sec) typically is preferable to reduce deposition and hydrogen sulfide generation.

On steep grades, when velocities are greater than 3 m/s (10 ft/sec), care must be given to the selection of pipe materials to prevent pipe wall erosion. At these velocities, the designer should consider using a drop manhole concept to reduce the flow velocities. Also, special attention to anchoring to prevent the pipe from creeping downhill should be installed. Anchors or retards should be coordinated with the appropriate backfill material. For instance, as previously mentioned, granular backfill in steep trenches may shift, causing the pipe to pull apart, particularly if percolating water velocities scour the backfill fines.

Collection systems are designed for future peak flow. For this reason, problems may be encountered in minimum grade pipelines during the early years of their design life. Low initial flows that do not approach design capacity can create maintenance problems and may require frequent line flushing and cleaning.

2.7.1.5 Inflow and Infiltration Allowance

Most collection systems do not remain as watertight as they were on the day of their installation and performance test. After a new collection system is installed, it eventually develops defects or avenues for extraneous water to enter. Often, the older the system is, the greater the I/I; however, newer systems can sometimes exhibit as much or more I/I, depending on the quality of construction and local conditions.

Design allowances for I/I are best developed for new pipe and collection systems from flow monitoring data collected on the existing sewer system or by benchmarking against other similar collection systems. Allowances for I/I should be coordinated with design standards that result from system conveyance planning design rates previously discussed.

2.7.2 Pressure Pipelines

2.7.2.1 Types of Pressure Systems

Pressure systems can be either positive or negative. Most systems are positive-pressure, whereby a pump transfers energy to lift the wastewater to the desired elevation at a specified rate. The most common positive-pressure system results when a pumping system receives wastewater from a gravity wastewater or other pumping systems. In a pressure wastewater system, the pipeline is called a force main. Force main diameters for these systems typically are 10 cm (4 in.) or greater. System pressures typically range from a few meters to approximately 60 m (200 ft) of total dynamic head, with the norm being between 15 and 30 m (50 and 100 ft) of total dynamic head.

Another form of positive-pressure system is a low-pressure pumping system wherein a pump receives flow from a single facility or a small group of residential or commercial facilities. The type of low-pressure pump used is influenced by whether the pump receives the wastewater directly into the sump pump or from an existing settling compartment such as a septic tank. The designer has the basic option of using a centrifugal or positive-displacement pump. Residential systems that use grinder pumps at residence are also common in many parts of the country.

The pressures produced typically are less than 46 m (150 ft), and the discharge flows to a common lift station wet well or a gravity system. Pipeline diameters from low-pressure pumps are typically 4 cm (1.5 in.) and pump to a 5- to 10-cm (2- to 4-in.) diameter header pipe. These systems can be used where geographical or topographical features would make a gravity system cost-prohibitive, such as in areas along lakes, bayous, or other waterfronts that either have high groundwater or variable terrain elevations. Low-pressure systems also are used in areas of rocky terrain or where population density is sparse.

Infiltration and inflow into both traditional and low-pressure pumping systems is relatively minor compared to gravity sewer systems. Essentially, the only component of the pressure system that is subject to I/I entry is the sump or pump wet well.

A negative-pressure system (also referred to as vacuum sewer) produces a negative pressure within the pipeline between the residential or commercial structure and the pump. The wastewater is drawn to the lift station where it discharges to a gravity sewer or another pumping system. A negative-pressure system typically has a pipeline diameter of 5 to 20 cm (2 to 8 in.). Such systems have the same applications as the low-pressure systems previously discussed. Negative-pressure systems also are subject to minor I/I because of the potential for leaks. For additional detail on systems such as pressure, vacuum, and small-diameter gravity sewers, see *Alternative Sewer Systems* (WEF, 2008).

2.7.2.2 Sizing Considerations

Most piping systems function as a single-barrel pipeline; however, a dual or parallel pipeline is not uncommon where service redundancy is extremely important or where clearance conflicts might prevent a single-barrel pipeline from being installed. In either situation, sufficient internal minimum velocities of 0.61 m/s (2 ft/sec) and, preferably, 0.91 m/s (3 ft/sec) must be reached to re-suspend organic and small, sand-sized solids that enter the pipeline. As with gravity sewers, the early years of a new installation may see flows that are at the low end of the design range. Therefore, the designer should evaluate whether minimum velocities will be reached daily. If not, then it may be necessary to develop an O&M plan to create enough flow to flush the pipeline. The extent of the flushing plan will depend on length of the pipeline. As an alternative, it may be necessary to include a pipeline pig launcher and retrieval appurtenances in the system design. A pig is a plug-type insert in the pipeline that uses hydraulic pressure to mechanically scrape or dislodge material from the pipe.

Maximum velocities for wastewater positive-pressure systems should not exceed 3 m/s (8 ft/sec) for normal, sustained flows. Wastewater conveys sand and grit that accelerates the inside deterioration or erosion of pipe coatings or wall thickness at bends or valves. Velocities from 0.9 to 1.8 m/s (3 to 6 ft/sec) are more desirable for sustained flows. The ability to stay within these ranges will depend on the flow range that the pipeline must convey.

2.7.2.3 Materials

Typical pipe materials commercially available for pressure sewer construction include

- Fiberglass reinforced plastic,
- Ductile iron,
- Prestressed concrete cylinder pipe,
- Polyvinyl chloride,
- High-density polyethylene, and
- Steel.

The same evaluation considerations apply to pressure pipe materials as previously discussed in the selection for gravity pipe materials. However, there are additional O&M considerations when evaluating pressure pipes. Pressure pipes tend to be installed at shallower depths than gravity sewer pipes. In colder regions of the country, pressure pipes should be installed below the frost line. Therefore, native soil character, surface contaminants that percolate through the soil, and stray currents from other protected structures can influence the exterior corrosion rate of the pressure pipe. Interior corrosion can be more aggressive in pressure pipes at high profile points and at the discharge to the atmosphere.

Pressure pipes, particularly nonmetallic, should be installed with a system for recovering the location of the installed pipe. This can be done with accurate record drawings or by burying a wire, tape, or electric signal coupon above the pipe.

2.7.2.4 Backfill

Backfill has the same important structural support function as discussed for gravity sewer pipe. Soil conditions can vary widely in a project, and it is typical to use local, excavated soil as backfill around the pipe. Geotechnical engineers should assist in closely evaluating soil characteristics, which influence the performance of thrust restraint for pressure systems.

Earth loads typically are not as great an external load because of the shallow depths sometimes involved. However, live loads can become significant. Designers should thoroughly review the pipe manufacturer's backfill design suggestions when designing the pipe and bedding combination.

2.7.2.5 Restrained Joints

The design of thrust restraint for pressurized, buried pipelines with push-on-type joints is important because of the damage that can result from joint separation. Pipe systems that have all joints restrained, either as furnished by the pipe manufacturer or from a welded joint, may not require external thrust restraint.

Thrust occurs when pipeline size or direction changes. Thrust forces can be hydrostatic, hydrodynamic, or thermal in origin; designers are responsible for evaluating each type of force. The designer should review the pipe manufacturer's recommendations for restrained joints and determine if application of the restrained pipe length and method is appropriate for the proposed site conditions.

Resistance to thrust forces can be provided by concrete thrust blocking, concrete thrust collars, or anchorage to a structure, or, indirectly, by transmitting the forces to the adjacent straight pipe sections using joints designed to resist shear and separation. The forces in the adjacent pipe are then transmitted to the soil by a combination of friction and lateral pressure.

Concrete thrust blocks depend on the allowable lateral soil pressure to develop the necessary resistive force. The allowable lateral soil pressure for local conditions must be determined during design and properly observed during construction. In adequate soils and for smaller diameter pipe, thrust blocks typically are the most economical method of resisting thrust. However, sometimes soils are unable to resist lateral force or dead weight force of the block. Also, there are times when space limitations or the probability of future excavations will dictate the need for an alternative restraint method.

Designers should also check for corrosion requirements to determine what protection is necessary for the thrust-restraint system.

2.7.2.6 Blowoff and Air Valves

There are several types of air valves that are necessary for proper operation of a force main: air release, vacuum, and combination air/vacuum. Air that enters or is entrapped in the pipe must be released. Because flowing water is subject to changing pressures and velocities, air is continually coming in and out of solution when the pipeline is in service. Unless air release valves are properly located, this free air will accumulate at high points within the system. Serious flow restrictions or even blockage can result from air pockets in the pipe, which can contain corrosive vapors and should be vented to reduce interior deterioration. Additionally, on long downhill segments of the pipe, a vacuum—negative pressure—can be created and cause a collapse if air is not allowed into the pipe.

Surge-pressure forces will be generated in the pipeline during situations such as pump startup, rapid shutdown, or sudden valve closure. The forces created can be substantial and need to be evaluated to determine their influence on the design. Whether surge suppression is accomplished through a surge-relief tank or air release valves, provisions for their regular maintenance should be established.

Valve maintenance is especially important on wastewater pipelines. Corrosion-resistant materials for the entire assembly often are overlooked. Use of corrosion-resistive materials should be considered throughout the system from the corporation stop through the shutoff valve and air valve internal parts. Otherwise, when a leak develops, maintenance personnel might not be able to close the shutoff valve or the corporation stop.

The designer needs to ensure correct sizing and selection of the air valve. Technical bulletins published by the manufacturer or included in reference materials will have recommendations for making the correct selection. These materials also will include recommendations for equipment and procedures for periodic valve cleaning. The designer should take into consideration these maintenance requirements and provide adequate access space for a maintenance worker to repair or remove a valve. This could include providing proper vault drainage to prevent contamination of adjacent surface water or groundwater and including provisions for odor control if the valves are close to residential or commercial development.

Similarly, sediment and debris can accumulate in low points; methods to remove these items need to be provided during design. Consideration should be given to avoid low points; when needed, blowoff valves should be provided at low points.

2.7.3 Odor and Corrosion Control

Without proper consideration of odor and corrosion control, the wastewater conveyance O&M staff will encounter continuous maintenance and repair problems from a deteriorating piping system as well as complaints from local businesses and homeowners.

Odor control may be accomplished by many different methods, including reducing turbulence in the piping system, using activated carbon on air valves, or adding chemicals and biofilters at lift stations. The designer should understand potential odor issues and take measures to control odor as required.

This section will not discuss corrosion control methods in detail but will outline important considerations in the selection of steel, iron, and concrete pipe materials for wastewater conveyance applications because these are the most susceptible to corrosion. A corrosion engineer or manufacturer representative should be consulted for specific corrosion protection controls. In addition, the designer should not overlook metallic components (such as fittings, bolts, bands, and valves) that are used with PVC, fiberglass, and HDPE conveyance systems.

Corrosion is a result of the reaction between the pipe material and its environment. Corrosion can occur on the interior of the pipe as a result of chemical reactions with wastewater. The exterior can corrode because of reaction with the soil. Designers often fail to anticipate the corrosion potential of electrical wiring, switches, and fixtures. The factors that must be considered when evaluating materials are determined after field testing and evaluation of the environmental conditions.

Bonded pipe coating and linings help control corrosion. They isolate the pipeline from its environment. However, coatings or linings are not perfect, and additional techniques may be required to control corrosion. Portland cement concrete or mortar can also be used to protect steel in non-detrimental environments. This is the basis for protection of the steel components of concrete pressure pipe and for mortar-coated and lined steel and iron pipe.

Polyethylene encasement is an unbonded coating traditionally used to protect the exterior of ductile iron pipe and fittings. This system uses polyethylene sheets that are loosely wrapped around the pipe and fittings. This barrier reduces the exterior corrosion by reducing exposure to oxygen, corrosive species, and stray currents. Polyethylene encasement may provide economical control from external corrosion in certain environments. However, it is important that the polyethylene encasement system be installed correctly to provide the necessary protection.

Cathodic protection is useful for mitigating external corrosion of both metallic pipelines and concrete pressure pipe. Cathodic protection can be an external source of direct current voltage applied to the pipe so that the pipe becomes lower in electrochemical potential than the soil around it. Thus, the pipe becomes the cathode rather than an anode. Sacrificial anodes such as magnesium or zinc rods can also be attached to the metal pipeline and buried in the soil around the pipe. The magnesium or zinc rods corrode sooner than the metal pipe because the electrochemical force developed between the metals causes the anode to lose the metal ions before the pipe does.

Application of cathodic protection to a coated metallic pipeline supplements the protection already afforded by coating systems. Cathodic protection systems can be quite complex and should be designed by corrosion specialists trained in assessing when and how to implement this method of control. Cathodically protected systems require regular O&M to read and monitor the status of the protection system.

2.7.3.1 Sulfide Control Design Considerations

The following guidelines have been prepared to help minimize the production of dissolved sulfide and the release of hydrogen sulfide gas into wastewater collection systems.

Sulfide control guidelines for gravity sewers are as follows.

- Sufficient velocities should be maintained in gravity sewers to minimize the production of dissolved sulfide as described previously.
- Excessive velocities that would cause turbulence of the wastewater and stripping of hydrogen sulfide gas should be avoided. Flow velocities should not exceed limits previously discussed. The sulfide concentration in the wastewater to be conveyed by the gravity sewer should be considered before designing for velocities that could cause the stripping of hydrogen sulfide gas.
- Turbulence at points of intercepted flow (manholes or wyes) should be minimized through the use of smooth transitions. Hydraulic drops and outfalls should be minimized. The use of drop manholes should be minimized. Optional methods such as submerged-outlet drop pipes or contained drops should be considered for drops of up to 6 m (20 ft) to contain and minimize turbulence.
- Combining wastewater flows at a confluence angle greater than 90 deg should be avoided. Confluence angles of 45 deg or less have been found to reduce wastewater pressurization, turbulence, and corrosion.
- It is recommended that wastewater pipelines not be installed with opposing flows (confluence angle of 180 deg).
- The use of inverted siphons should be avoided whenever possible. Inverted siphons should be provided with adequate air jumper capabilities and positive corrosion protection measures such as use of liners or inert materials.
- It is recommended that diameter changes only occur at manholes that have been properly designed to allow for change in flow characteristics.

Sulfide control guidelines for lift stations and force mains are as follows.

- Lift stations and force mains should be designed to minimize hydraulic detention time within the pipeline, thereby reducing dissolved sulfide production.
- The discharge of force mains into the wet well should be designed to minimize turbulence and the stripping of hydrogen sulfide gas. Wherever possible, discharges should be submerged to absorb hydraulic energy and minimize turbulence.
- Gravity wastewater entrances into wet wells should be designed to minimize free fall and turbulence. Lift stations should be designed so that free air movement through the wastewater is not impeded by wet well water levels except at flows greater than peak dry weather flow.
- The number of air release points (placed at high points) for force mains should be minimized through grade and slope control. Drainage for the force main should be provided if possible. To avoid localized corrosion, partially full force-main conditions should be avoided at any point in the pipe. Force mains should be designed with a flat or continuously rising profile whenever possible. If possible, force mains should not be designed with a downward or negative slope. The idea is to control grade as much as possible and place air release valves as needed.
- Dual-force-main piping should be considered. This will help maintain minimum velocities, provide for future capacity, and minimize dissolved sulfide generation (if force mains can be drained).
- The lift station could be designed so that pumps clean the wet well. Self-cleaning wet wells may be advantageous for wastewater.
- Where the force-main profile cannot be controlled and air releases must be used, the use of variable frequency drives on pump motors should be considered to maintain flow conditions and avoid excessive operation of the valves.

2.7.3.2 Corrosion Protection Design Considerations

The following guidelines have been prepared to reduce corrosion potential in wastewater collection and treatment systems caused by exposure to hydrogen sulfide gas. In general, hydrogen sulfide gas concentrations less than 2.0 ppm are not considered corrosive. Hydrogen sulfide gas concentrations between 2.0 and 5.0 ppm are considered potentially corrosive, and concentrations greater than 5.0 ppm average justify protection against corrosion.

Corrosion protection guidelines for gravity sewers are as follows.

- Pipe material and pipe liners used for wastewater collection system construction should be unaffected by contact with sulfuric acid solutions as low as pH 0.5 (approximately 7% sulfuric acid by weight).
- Cementitious concrete products of any type (calcium silicate, calcium aluminate) should not be used for wastewater conveyance wastewater without an appropriately designed corrosion protection system (liner).
- Particular attention should be given to corrosion prevention of existing, unprotected concrete gravity sewer pipes and manholes in areas of increased turbulence (intercepted flows, 90-deg bends, slope changes, drop manholes, and force-main discharge locations) for a distance of 300 diameters downstream and 50 diameters upstream.

Corrosion protection guidelines for lift stations and force mains are as follows.

- Lift station wet wells should be constructed of or lined with inert materials where hydrogen sulfide is anticipateds. Exposed, unprotected concrete of any kind should not be used in lift station wet wells or other areas exposed to average concentrations of hydrogen sulfide gas greater than 5.0 ppm.
- Piping used inside lift station wet wells or areas of treatment facilities exposed to significant concentrations of hydrogen sulfide should be constructed of or lined with inert materials. The recommended metal is 316 stainless steel. Coated ductile iron is also not advisable.
- Force mains should be constructed of inert pipe materials. Prestressed concrete cylinder pipe and ductile iron pipe are not typical for wastewater force-main applications, as additional corrosion protection is required for these pipe materials. There are linings available for prestressed concrete pipe and ductile iron pipe that enhance their corrosion resistance.
- Particular attention should be given to protection of structures downstream of turbulent force main discharge areas. It is even better to avoid designs that cause turbulence.
- Force main discharge manholes should be constructed from inert materials or protected with materials such as a PVC liners or epoxy coating.

2.8 Rehabilitation

Rehabilitation maintains the structural and operational integrity of the collection system. The methods available to evaluate and rehabilitate the collection system have become more sophisticated with the advancement of robotics and improved camera systems. *Existing Sewer Evaluation and Rehabilitation* discusses this subject in more detail (WEF & ASCE, 2008).

Both gravity and pressure pipelines will require repairs or replacement during its lifespan. Many of the factors that create the need to repair or replace pipe have been discussed. Typically, the primary reason for repairing or replacing conveyance system pipe is related to structural or I/I problems.

2.8.1 Condition Assessment for Design

The most critical part of any rehabilitation program is to understand the pipe condition and need for repair or rehabilitation. Wastewater system condition assessment technologies such as closed-circuit television (CCTV) and smoke testing are discussed in more detail in Chapter 4 of this manual. Many improvements have been made over the last 10 years in condition assessment techniques and methodology. Regardless of the method used, it is best when the designer and the system owner speak the same language.

2.8.2 Wastewater System Assessment Protocols

In 2013, the National Association of Sewer Service Companies (NASSCO) developed wastewater system condition assessment protocols called the Pipeline Assessment and Certification Program and the Manhole Assessment and Certification Program. These programs were developed to promote the use of common standards within the wastewater industry including procedures, a coding system, and a consistent method of assigning severity codes to identify, quantify, and prioritize defects in an existing collection system. For further information on this program, please visit http://www.NASSCO.org.

2.8.3 Structural Repairs

The need to make structural repairs typically becomes evident either through a sudden pipe failure, through field inspections such as internal CCTV inspections, or pipe-wall coupon sampling. In each case, an exact cause of the problem needs to be determined so that the corrective action is effective. Otherwise, the corrective action may have to be significantly overdesigned to account for imprecise pipeline assessment and thus account for all of the possible or speculative causes of the problem.

Structural repairs typically consist of replacing the section of piping system with the same material or an improved more modern material with manholes or material adapter couplings at the change of material locations. It would be advisable to use a granular pipe-bedding material to provide stable, uniform support for these repair locations.

The designer also should consider coordinating with the pipe manufacturer to discuss their recommended replacement couplings and other installation recommendations. If a section of pipe is replaced, the internal pipe diameter of the spliced-in pipe should match the internal diameter of the adjoining sections. Otherwise, the joint between the old and new pipe could become a point where solids and debris become lodged.

2.8.4 Technologies for Design

Repairs for both pressure and gravity pipelines also can be made to the exterior using clamps or sleeves or to the interior using a liner either for long stretches or for several meters. For longer pipe segments, rehabilitation systems include slip-lining, swage-lining and rolled-down pipes, segmented linings, cured-in-place (CIPP) systems, fold-and-form pipe (FFP) systems, spray liners with geoweb, and spiral wound pipe. The systems have become increasingly competitive and beneficial because of their less-disruptive trenchless technology features. CIPP, FFP, spray liners, and spiral wound pipe systems have applications for structural point repairs as well. It is important to weigh the ongoing O&M cost associated with a structural point repair versus a manhole-to-manhole full pipe replacement. The O&M costs may include SSOs and associated fines and emergency repairs down the road as more pipe fails. If possible an assessment of the damaged line should be performed to help educate the owner on the status of the system in the immediate area of the proposed point repair.

A critical principle to consider in structural design of the rehabilitation technologies available is that the structural load the liner must support is determined by the degree of degradation of the original pipe. Design conditions for structural liners typically fall into two primary categories: partially deteriorated and fully deteriorated. There are many materials and systems for both external and internal rehabilitation that are commercially available. These materials and systems are often advertised in national trade organization publications.

Pipe bursting or pipe eating are other forms of pipe replacement. These systems break the original host pipe, and a new HDPE or fused-PVC (FPVC) pipe is pulled into its place. Pipe upsizing of one to two diameters is possible under certain conditions. Both pipe bursting and liner systems require that individual service connections be reestablished either by using robotics inside of the pipe or by excavating to expose and reconnect the service connection.

Please refer to *Existing Sewer Evaluation and Rehabilitation* for more data on pipeline rehabilitation means and methods (WEF & ASCE, 2008).

2.8.5 Inflow and Infiltration

Inflow and infiltration is a problem associated with gravity piping systems. The term refers to extraneous flow that enters underground wastewater pipes and imposes an undesirable hydraulic load on the collection system. Infiltration sources include groundwater, which may be prevalent year-round, and rainfall, which is present during and shortly after rain events. Inflow enters the pipeline via direct connections to the collection system such as through manhole covers, surface or yard drains, roof downspouts, sump pumps, or service connection cleanouts with missing caps. Storm water exiting the drainage system through damaged pipes, non-water-tight joints, and unsealed lift holes may follow nearby collection system trenches and enter through damaged pipe and non-water-tight joints.

Rehabilitation approaches to correct infiltration may be different than those to correct structurally damaged pipe. Television inspection of the pipeline can help define the extent of structural rehabilitation needed. Groundwater migration, however, will influence the required rehabilitation to control infiltration. Therefore, infiltration rehabilitation work may include sealing or making each pipeline and service connection joint watertight the entire length of the pipeline within a groundwater zone of influence. Rehabilitation might include chemically grouting each joint (or at least pressure testing each joint) or relining the original pipeline with one of the many commercially available lining materials discussed previously.

2.9 Manholes

Manholes are installed in collection systems to provide access for routine or emergency maintenance, as well as to meet regulatory requirements regarding grade or alignment changes. They are sometimes rendered inaccessible because of location, negligence (for example, covered by pavement), or design deficiencies. If manholes are covered during an overlay project, it can be contentious as to who pays for the repairs necessary to access the manhole and subsequently raise the lid. Locating wastewater lines in street or alley rights-of-way is preferable to the use of easements, where obstructions such as utility poles, meters, junction boxes, fences, retaining walls, trees, shrubs, or utility buildings can restrict access. It is important to avoid placing manholes and pipelines in the path of direct, continuous traffic; it is best to locate these items outside wheel areas.

Manholes typically are placed at each change in direction horizontally and vertically and change in pipe diameter so that pipelines are straight between manholes as a result of several considerations: straight pipeline layout and construction is easier; accurate record drawings have often not been kept to allow relocation of curved segments; and pulling CCTV and cleaning equipment in curved pipelines is more prone to operational problems and pipe damage. Curved pipelines are more prevalent in older systems even though a properly designed curved pipeline will convey wastewater just as effectively as straight pipelines and can cost less. Some communities, however, have adopted the practice of installing a manhole immediately upstream, downstream, or at both ends of the curved portion of the pipeline if additional construction work is done in the vicinity of a curved collection system.

Ideally, manholes are large enough to provide space for a person to work with cleaning equipment, a shelf (also called a bench) wide enough to provide safe footing, and support for a ladder. Manhole walls can be sealed from groundwater infiltration by using mortared joints, adequate flexible sealer, or rubber Oring gaskets for precast section manholes. It is reasonable to consider a combination of joint-sealing methods because concrete casting tolerances may not produce tight-fitting tongue-and-groove precast joints.

Manholes typically have an inside diameter of at least 1.2 m (4 ft) but should be larger when the manhole is greater than 3 m (10 ft) deep to help with access and maintenance. Larger pipelines need correspondingly larger manholes to provide an adequate shelf on which to work and a wall surface sufficient for a solid, watertight connection. Eccentric manhole cones (i.e., cone is offset to be aligned vertically with manhole wall) are easier and less dangerous to enter. To reduce problems with some maintenance equipment, the straight side is sometimes over the outlet pipe. Many agencies require construction of concentric manholes (i.e., entrance cone is in the center of a manhole), which facilitates the insertion of cleaning equipment in multiple directions regardless of orientation. Because of safety considerations, some agencies have eliminated manhole steps and now use ladders or hoists.

2.9.1 Spacing

Manholes typically are spaced at intervals of 120 to 150 m (400 to 500 ft) for small sanitary sewers. Spacing often is specified by local regulation. With the advent of modern cleaning equipment, some agencies have considered lengthening this maximum interval. Although some new cleaning equipment is able to clean up to approximately 300 m (1000 ft), other maintenance activities, such as bypassing a section or certain rehabilitation activities, can be difficult or impossible at that distance. Where there are truck-access limitations (such as in a backyard or parkway), the distance from the access point to the manhole needs to be considered in determining the manhole spacing. Intervals of 150 to 300 m (500 to 1000 ft) are typical for large sanitary sewers of 0.75 m (30 in.) diameter or greater.

A manhole is also necessary at each change of grade, invert elevation, or pipe size; at each intersection of two or more pipelines; at the terminal end of each pipeline; and at major service connections (such as industrial facilities and schools). Manholes should not be placed within street intersections.

The practice of replacing the last manhole with a lamphole or vertical riser is not recommended. When collection systems must be located along a rear property line, the pipeline should be extended to a street right-of-way or an access easement, even if this requires construction of an additional dry segment. This ensures access for the maintenance crews.

2.9.2 Materials

2.9.2.1 New Construction

Most new manholes are made of precast concrete. Other materials such as fiberglass, polymer concrete, and HDPE also are used in environments that are detrimental to concrete. Some communities require coal tar epoxy, elastomeric epoxy, or PVC liners in the manholes that are to be located on interceptors greater than a specified diameter. A newer technology is antimicrobial additives that are included in the mix of the concrete, protecting the concrete manhole from corrosion while eliminating many of the steps necessary for a liner to be applied. The walls of new manholes typically are heavily pitted and need to be filled with a cement or nonshrink grout layer before application of an epoxy liner. Otherwise, corrosive gases will penetrate existing pinholes, corrode the area behind the liner, and eventually delaminate the liner. Precast concrete manholes that are located in the upper reaches of collection systems or where hydrogen sulfide generation is minimal will provide years of good service.

2.9.2.2 Rehabilitation

Manhole rehabilitation typically begins with investigations to help identify I/I problems in the collection system. Many communities prefer to rehabilitate not only their I/I defects but also their structural defects under the same effort. Therefore, a combination of structural, waterproofing, or corrosion control methods may be necessary at the same manhole. Some manufacturers advertise products that meet all three requirements. The broad categories of manhole rehabilitation are chemical grouting, cementitious manhole restoration, polymers (epoxy, polyurethane, and urea), cured-in-place liners, chimney seals, barrel joint seals, and precast inserts. See *NASSCO Performance Specification Guidelines for Renovation of Manhole Structures* for more detailed information (http://www.NASSCO.org).

2.9.3 Channels

Smooth, continuous manhole channels efficiently move wastewater from one or more inlet lines to the outlet lines. Properly designed and constructed side or angle inlets prevent deposition of solids at the transition point.

Wherever a change in pipe diameter occurs, the pipe crowns, or water surfaces are located at the same elevation to reduce flow resistance, check with the system owner and local regulatory agency to confirm requirements. Where there is a change in pipe direction without a change in diameter at a manhole, a drop in elevation is allowed through the manhole. The following drops are recommended: a 0- to 45-deg angle change over a distance of 3 cm (0.1 ft) and a 46- to 90-deg angle change over a distance of 6 cm (0.2 ft). Check with the system owner and local regulatory agency to confirm requirements.

Inlet pipes that are too high to enter the manhole at the bottom, more than 31 cm (1 ft) can enter through a drop connection. However, every effort should be made to avoid sudden drops in flow. Sudden changes in flow direction will cause gases such as hydrogen sulfide to be released out of solution and can cause odor issues.

Constructing shelf areas so that they drain to the channel at a slope of at least 1:12 reduces deposition. Because of careless construction, a channel of good hydraulic properties, though important, is not frequently created. The channel should be, as far as possible, a smooth continuation of the pipe. To this end, the pipe is sometimes laid through the manhole with the top half removed to provide the channel (ASCE & WEF, 2007). The channel walls should be formed or shaped to three-fourths the full height of the crown along interceptors and other manholes with high velocities. This practice will help reduce deposition of solids on the shelf and facilitate smooth flow hydraulics through the manhole. Making the channel wall the full height for most 0.2-m (8-in.) diameter sewers increases the difficulty of wastewater investigations and maintenance tasks such as

- Physical inspection of pipes by lamping,
- Insertion of television inspection cameras,
- Deployment of cleaning equipment,
- Insertion of plugs,
- Installment of some flow measurement equipment, and
- Full height shelfs or benches are beneficial as they keep full pipe flow from hitting the walls when moving through the manholes.

2.9.4 Covers

Manhole frames and covers are constructed of heavy cast iron, composite materials, or ductile iron. They need to be approximately 76 cm (30 in.) in diameter to allow easy access and meet regulatory requirements. Machining the horizontal bearing surfaces between the cover and frame reduces rocking and provides a tighter fit, especially if a rubber gasket is grooved into the frame or lid to be watertight.

In flood plains, floodprone areas, or areas subjected to ponding or sheet flow runoff, manhole covers should be watertight (with or without bolts). These covers and rubber gaskets are designed to eliminate inflow. Lifting holes or bars in the lid are designed to be non-penetrating to reduce surface water inflow. Removal methods that do not require lifting holes are preferred; however, if lifting holes are necessary, there should be no more than two. Alternatives to replacing an existing hole-penetrating cover are either hole plugs or frame and cover inflow inserts (dish). The inserts stop or impede the inflow from entering the system through the cover penetrations. The inserts are made of plastic, polyethylene, or metal materials. Check with your local system operators, as inserts are not universally accepted.

Where manholes are located in poor soils, shifting sands, or soft asphalt pavement on a hot day, it is common to install a concrete ring that extends approximately 0.3 m (1 ft) around the top of the manhole frame and cover. This provides maintenance crews with firm footing for a confined space tripod and for lifting bars. A secondary benefit of the concrete ring is that it makes the manhole easier to locate both in drive-by searches and in aerial photographs.

Adjustment rings between the cone or slab and the rim casting may be appropriate in manholes located in streets and alleys. These rings facilitate adjustment for regrading projects. It is important to take care to waterproof and seal the joints between the manhole frame, adjustment rings, and cone because this is often the area where inflow enters the manhole.

In cold climates or where traffic loads are heavy, adjustment rings may fail and allow I/I or loss of backfill. Rings of highquality concrete may be used to help avoid this. Flexible frametocone seals that permit movement from frost action or traffic effects are a viable alternative.

2.9.5 Steps

Corroded or poorly grouted manhole steps (rungs) present a safety hazard. These steps can break off underfoot, possibly causing a serious or fatal fall. Efforts should be made to remove steps as part of a rehabilitation program. Most agencies are dispensing with steps in new manholes and equipping maintenance vehicles with lightweight aluminum or fiber-glass extension ladders or portable hoisting frames and harnesses.

2.9.6 Drop Manholes

Drop manholes are manholes in which the vertical distance between the influent pipe and the effluent pipe is greater than 0.3 m (1 ft). Drop manholes that have a drop greater than approximately 0.6 m (2 ft) should have two influent pipes. The top pipe enters the manhole on the same grade as the upstream pipe. The bottom pipe has the same vertical alignment as the upper pipe but comes into the manhole at the invert of the manhole. There are both outside and inside drop manholes used in collection systems. The design again depends on the jurisdiction and their experiences or preferences.

Drop manholes typically are considered when it is not cost effective to make an influent line match the effluent line. An example of this is when the construction cost savings of installing the pipe at the extra depth exceeds the cost of the extra piping and workmanship of the drop connection at the manhole. However, because drop manholes can create other costs associated with material corrosion and maintenance problems, other approaches are often used. For more information, see *Gravity Sanitary Sewer Design and Construction* (ASCE & WEF, 2007).

2.9.7 Connection Between Manhole and Wastewater Pipeline

Design should anticipate the possibility of differential settling of the manhole and the wastewater pipeline. Design and construction precautions are warranted where the manhole and pipeline connect. Shear cracks or stresses in a pipeline can occur at or immediately outside the manhole. The use of flexible pipe-seal connections integral to the manhole wall is a method typically specified in design to avoid buildup of stress forces in the pipe. If flexible pipe-seal connections are not applicable or used, a flexible pipe joint at, or within approximately 1 m (3 ft) of the outside manhole wall, decreases the likelihood of pipe damage.

Designers should allow for the change in trench width as the pipe enters and leaves the manhole. Also, the contractor should properly place, compact, and support the pipe with an appropriately graded granular bedding and backfill material. Otherwise, the pipeline may develop shear stresses and fail.

Corrosion problems should be expected where a force main discharges to a manhole as discussed previously. Hydrogen sulfide (generated principally by the bacterial reduction of sulfate in wastewater) is released, oxidized, and condensed on exposed surfaces as sulfuric acid. This process can cause extreme damage to cement, carbonaceous aggregate, asbestos-cement, unlined ductile iron, and steel. Limiting turbulence at these junction points will reduce the release of hydrogen sulfide gas.

2.10 Manhole and Pipeline Testing

2.10.1 New Construction

Emphasis on eliminating I/I from the collection system has spurred the enforcement of performance tests that allow only relatively minor amounts of infiltration or exfiltration. Hydrostatic tests typically are used for pressure pipelines, manholes, and, to a lesser extent, small-diameter gravity pipelines. Pressures for pipelines are established based on design requirements of the pipeline system and should represent the greatest pressure to be generated in the pipe. Pressures for manholes are static pressure only; the manhole is filled with water and the rate of exfiltration is measured.

Air testing typically is used for gravity pipelines and manholes. However, manhole pressure tests are negative-pressure or vacuum at approximately 14 to 21 kPa (2 to 3 psi). Pressure increases over time are measured against acceptable rates.

In addition to air or hydraulic tests, it is a good practice to pull a mandrel through the installed pipe and to require CCTV inspection to note any observed leakage or defects. Mandrels are used to determine whether an installed pipe has deflected more than accepted by the designer. Mandrel tests typically are used with flexible pipe installations because they will deflect easily if not properly backfilled. Two basic types of mandrels are available. One is a rigid mandrel constructed with a fixed deflection that, if exceeded, will hedge in the pipe and not advance in the pulled direction. The other is a mandrel with telescoping prongs that are depressed by a deflected pipe and, when removed from the pipe, are measured against the designer's allowable deflection criteria.

Excessive deflection reduces flow capacity, can cause structural stress points on the pipeline, and can worsen over time. Television inspection for new construction can reveal manhole and pipe leaks or defects that may exist but that do not cause an air or hydraulic test failure. Also, CCTV can reveal line sags by adding a small amount of dye to clear water released into the new pipe. Closed-circuit television can also reveal sediment or debris left in the pipe after construction.

Some communities require each manhole and pipe to be tested or CCTV inspected. Others establish a minimum percent to be tested or televised with provisions for more if specified criteria are not met. Using CCTV to inspect the entire system before acceptance (and before paving above the pipe) not only confirms that the system is installed correctly, but provides documentation for future comparisons.

2.10.2 Rehabilitation

The same tests for new pipelines and manholes apply to rehabilitated pipelines and manholes. However, rehabilitation of manholes should include additional tests typically not required for new construction (except for the previous cases). These tests are a surface density test (hammer test), an adhesion test, and a pH test.

Surface preparation for manhole rehabilitation is a key factor in achieving adhesion of the structural or corrosion-resistant liner to the manhole substrate surface. Surface density tests, pH tests, and petrographic tests (ASTM C856-04) can provide important information about the condition of the substrate material (ASTM International, 2004). Petrographic tests performed by a qualified person can reveal whether the concrete composition and the surface are cleaned properly. The adhesion test is an indicator test that ensures that the complete liner is properly applied.

2.11 Inverted Siphons

An inverted siphon is a section of sewer that, because of a large obstruction such as a river, depressed freeway, subway, or utility line, must be installed below the hydraulic gradient and thus operates under positive pressure. Most inverted (or depressed) siphons contain at least two pipes (barrels) although three pipes are preferred: a smaller pipe for low flows to provide greater velocity for scouring, a midsize pipe for larger flows, and a larger barrel to carry peak or future flows. To eliminate overflow or backup conditions, some inverted siphons include another pipe capable of carrying the entire flow in an emergency.

Inverted siphons typically have maintenance access structures at each end. Gates or stop plates at the inlet structure allow the diversion of all flow through any one barrel for scouring. Flap gates held open by a chain and accessible from the top of the manhole have been used successfully in small siphons. Working space is allotted in the inlet structure to launch a cleaning pig and in the outlet structure for recovery. To allow passage of cleaning equipment, smooth curves of adequate radius (less than 22 deg) are preferred. Inverted siphons have been successfully constructed with various types of pipe, including ductile iron, concrete, FPVC and HDPE.

Odor problems tend to occur at the inlet chamber because air is released by movement induced by wastewater flow and lack of a free-water surface through the siphon. Positive pressure develops in the pipeline upstream of a siphon because of the downstream movement of air induced by the wastewater flow. Air tends to exhaust from the manhole structure at the siphon inlet and can cause odor problems. The problems typically are cured with air lines or air jumpers. Alternatively, you can look at passive or active odor control systems. These were discussed in Section 2.7.3 of this chapter.

The inlet chamber has a lateral overflow weir and, sometimes, a sump to trap grit and heavy objects and to prevent clogging of the barrels. Screens may be installed if the added expense of periodic cleaning is warranted. Inlet and outlet structures both need tamperproof manhole covers.

Venting is required on both chambers for the siphon. The inlet (upstream) chamber needs to be vented to relieve the positive pressure created by the siphon, and, conversely, the outlet (downstream) chamber needs to be vented to relieve the negative pressure created by the siphon. If venting is not provided, the siphon will air lock, increasing the head loss across the siphon resulting in poor hydraulics. The vents shall be sized based on the expect flows. Odor control is strongly recommended on both vents.

Both the inlet and outlet structures are prime environments for hydrogen sulfide release, which results in an aggressive corrosion attack on concrete and iron materials. Corrosion control is vital for these facilities if a reasonable life expectancy is desired.

2.12 Wastewater Pipelines Above Grade

Sanitary sewers are sometimes placed over an obstacle where topography, width of crossing, and flood elevations and velocities allow. They are installed or hung on bridges, supported on piers, suspended by cable spans, or fastened to a structural beam. These would be referred to as "aerial crossings." Pressure-type joints, insulation, and jacketing prevent leakage and protect the pipe against freezing, although above-grade pipelines are rarely used in colder climates.

2.13 Crossings and Tunnels

2.13.1 Stream Crossings

Collection systems that transect a stream and are installed using traditional excavation techniques often have top of pipe elevations that are several meters beneath waterway inverts. This typically is done by necessity because there is minimum slope available, forcing the pipe to be as shallow as possible. These crossings are subject to erosion and washout because of their proximity to the stream invert. Crossing pipe can be encased in concrete or placed in another pipe (casing) to protect against scour, impact, and potential inflow. Pipe joints designed specifically for crossing bodies of water also are available. When open cutting a stream, it is also crucial to consult with environmental scientists who specialize in streams and rivers to determine if the banks may widen or how much down cut is expected at a given crossing. If the crossing is complex, high risk, or expensive, alternative crossing options such as microtunneling or directional drilling should be evaluated. Regulatory agencies increasingly are requesting tunneling to avoid concerns with open-cut techniques.

Some agencies require that crossing signs be posted on either side of a navigable stream or lake that are visible from land and water. Signs should alert the public to the presence of the submerged pipe so that activities such as dredging, geotechnical explorations, and boat anchoring do not penetrate or shift the pipe's position and make it more vulnerable. A contact phone number and any other special notes should be legible from a reasonable distance as determined by the specific site conditions.

Substantial problems may be encountered where wastewater pipelines span a waterway, particularly when support piers are required. Brush and debris invariably jam against the pipe, sometimes causing a rupture and discharge of untreated wastewater to the stream. Bridged crossings ideally are placed above the maximum flood level, typically designed to the 100-year rain event. Stabilizing stream banks helps to reduce erosion maintenance. Manhole structures should be protected or physically located to reduce washout by bank erosion. Input from stormwater engineers and environmental scientists regarding where the stream is in its life cycle (i.e., widening or downcutting) is extremely important when performing construction near any type of waterway.

2.13.2 Other Crossings

There are many other cases in which pipelines need to cross major highways and busy intersections or to avoid public concerns. In these cases, tunnels or trenchless methods are being used more often. The method of tunneling depends primarily on the diameter of the crossing and the soil conditions.

When crossing major freeways, agencies will require an encasement pipe to provide additional protection to their facilities. This is especially true when using pressurized piping.

2.14 Service Connections and Disconnections

Service connections also are referred to as house connections, building sewers, service lines, or service laterals. They connect the customer's facility (building) to the public collection system. Service connection requirements for each community vary as to the criteria for design and installation.

Attention to service connections has greatly increased over the past 10 to 15 years because of the recognized difficulties with I/I. Because the length of service connection piping can equal or exceed the length of the public collection sewer system, their contributing effect on wet-weather flows can be significant. This fact recently prompted a change to the long-held cultural restraint of investing public money on private service connections; public money may be more effectively spent to remove defects. Regardless of the approach to correcting service connection defects, designers of wastewater pipelines should include realistic flow allowances for I/I from service connections.

Optimum collection system design includes properly located and securely plugged Y and T connections for future service connections. Typically, the public pipeline is installed and then the service Y or T connections installed with a section of service pipe plugged at the end before extending to the private property boundary. Close attention should be given to designing a tightly plugged pipeline; otherwise, the service connection can become a significant source of I/I and sediment. Inexpensive mechanical plugs are available. Also, accurate location records help make Y and T connections easily recoverable when the plumber wants to complete the building connection or perform repairs. Recently, cities have begun requiring developers in new areas to extend the public service lateral to the property line. A tee connection has a pipe extended above the proposed surface elevation to ensure the plumber is able to connect the private service line to the tee. The vertical pipe above the surface is commonly called a "city cleanout" for access to the collection system.

Sometimes, service connections are not available on the pipeline at the desired location. Service connection installation via the break-in or hammer-tap methods can cause pipe failure and line blockage and makes pipes vulnerable to infiltration and root intrusion. Local regulations typically prohibit break-in service connections and require machine-tapping and saddlestyle or other suitable factory-fabricated connections. It is strongly encouraged that construction inspection is performed on service laterals at the time of installation.

Old structures may be razed and the excavated land backfilled during property development. Typically, service lines are not properly disconnected, and the construction design will require a new service connection. Several generations of site reuse may result in multiple service connections. Inactive connections are a source of infiltration, insect and rat infestation, and other possible maintenance problems. Therefore, building permit and design regulations often may require that abandoned services be disconnected and permanently plugged at the main, if possible. Again, ensuring a watertight plug and an accurate location typically will restrict the potential problems that could otherwise result.

As stated above, more attention is being paid to service laterals and private system. Many communities are now ensuring that private systems are being rehabilitated as required to reduce the amount of I/I. One approach is requiring inspection and upgrades at the time of property sale or ownership transfer. Another is to have owners pay monthly into an insurance policy that can be used to rehabilitate laterals. Sharing the cost of lateral rehabilitation between the owner and the utility also has been used successfully in some communities.

A resource called *Private Property Infiltration and Inflow Control* (WEF, 2016) has been created to address ongoing developments in this area.

2.15 Value Engineering

Use of value engineering has typically been used only for large, multifaceted projects. The intent is to draw together experts in the areas the project addresses. For instance, a large collection system project with numerous tunnels and horizontal directional drilling (HDD) might have a team composed of a utility designer, a geotechnical engineer, a tunnel designer, an HDD specialist, a cost estimator, and certified leader. It is recommended to have a leader who has served on teams before and has had previous success in completing such projects.

Value engineering can also be accomplished by ensuring the end users of the system are involved as they understand some of the back-end maintenance issues that designers may not have considered. This may allow the designer to choose a less expensive route or may result in a more expensive initial capital cost but lower O&M costs over the life of the project.

2.16 Specification Writing

Specifying the type of materials that the project needs and their minimum performance criteria is important to project success. It is recommended that designers use the Construction Specifications Institute (CSI) standard format and guidelines when developing a set of technical specifications. The CSI is divided into many divisions that each deal with certain aspects of construction. Each division is further separated into specific sections dealing with particular items such as pipe material or testing. The format is straightforward and provides general information, references, descriptions of products and materials, and installation requirements and testing.

As part of specification writing, it is the responsibility of the owner and engineer to determine contract document language outlining the relationship between owner and contractor, as well as any involvement the engineer may have during construction. It is recommended that the engineer be involved in construction. Below are a list of project delivery methods. Chapter 6 of this manual discusses these delivery methods in more detail.

- Design–bid–build
- Design–build
- Construction manager at risk

Other options are available that involve third-party financing and O&M of the project post construction. The three listed above are the most common in the wastewater industry.

3.0 REFERENCES

American Society of Civil Engineers & Water Environment Federation. (2007). *Gravity sanitary sewer design and construction* (2nd ed., WEF Manual of Practice No. FD5/ASCE Manuals and Reports on Engineering Practice No. 60). American Society of Civil Engineers.

ASTM International. (2004). *Standard practice for petrographic examination of hardened concrete* (ASTM Standard C856-04).

Jones, G. M., Sanks, R. L., Tchobanoglous, G., & Bosserman II, B. E. (2008). *Pumping station design* (revised 3rd ed.). Butterworth-Heinemann.

National Association of Sewer Service Companies. (2013). *NASSCO performance specification guidelines for manhole rehabilitation.* https://www.nassco.org/sites/default/files/manholes.pdf

Water Environment Federation. (2008). *Alternative sewer systems* (2nd ed., WEF Manual of Practice No. FD12).

Water Environment Federation. (2016). *Private property infiltration and inflow control.*

Water Environment Federation & American Society of Civil Engineers. (2008). *Existing sewer evaluation and rehabilitation* (3rd ed., WEF Manual of Practice No. FD6/ASCE Manuals and Reports on Engineering Practice No. 62) McGraw Hill.

6

Construction Contracting

1.0 CONSTRUCTION PROJECT MANAGEMENT

This chapter provides an overview of public works construction contracting for wastewater collection systems and provides sources of more detailed information. Significant information and guidance also are available from the following organizations: Construction Specifications Institute (CSI); Engineers Joint Contract Documents Committee (EJCDC); Construction Industry Institute (CII); American Public Works Association (APWA); American Institute of Architects; Associated General Contractors of America; Associated Builders and Contractors; National Utility Contractors Association; United States Army Corps of Engineers; and American Society of Civil Engineers (ASCE).

Wastewater collection systems construction projects involve complex processes. Even small projects involve different skills, materials, equipment, and operations. The construction process is subject to variable and unpredictable factors. Many aspects of construction sites contribute to the complexity of construction projects: subsoil conditions, surface topography, weather conditions, transportation systems, material and equipment supply, availability of skilled labor, local subcontractors, and many others. Therefore, to some degree, every construction project is unique.

Construction drawings illustrate the extent and arrangement of the components of the facility. The specifications describe the materials and workmanship required. Construction contract documents are either prepared in-house or subcontracted to an independent consulting engineer, depending on the complexity of the project and the owner's design capabilities.

Project management is the overall control of design and construction to achieve an optimum balance between quality, schedule, and cost. Project management, however, does not exert a constant, uniform influence over these attributes throughout all phases of project delivery. Project management influence is greatest during the definition and preliminary planning phases, and tends to decrease rapidly as the process continues toward design and construction. Significant decisions concerning overall project size, scope, complexity, location, and performance characteristics are made during the early stages of the project. Following the early stages of the project, firm commitments of project resources must be made. As the facility owner commits resources to the project—such as acquiring facility sites, engaging consulting engineers for detailed design, and hiring contractors to perform construction—it becomes more difficult and expensive to make changes to the project plan. As a result, proper planning and decision-making are extremely important during project definition and preliminary planning.

Software is available to provide the tools and resources for project management. These project control systems include scheduling, estimating, and providing information management to control the budget, schedule, and quality of the project.

2.0 LAWS AND REGULATIONS

2.1 Contracting Requirements

There are several items to consider before a project is bid. Special laws such as building codes, environmental regulations, building permits, safety and health regulations, zoning regulations, prevailing wage requirements, and licensing for contractors may be applicable. These are intended to protect public health and safety and prevent fraudulent acts by owners and contractors. Contractors should be informed of these requirements during the bidding process.

Many states require all public works projects be designed by a licensed professional. Some states set a dollar limit, after which a licensed professional must prepare the design.

Public works managers need to identify and understand the various contracting requirements that are applicable to their state and particular jurisdiction, and their implications of the project.

2.2 Competitive Bidding

Laws or regulations typically mandate the competitive bidding of utility projects, ensuring value to ratepayers and fairness in the expenditure of public money. Public administration of bids typically is required for publicly funded work. Project bids, which are based on construction contract documents, are designed to receive cost-competitive bids and to provide the owner with a constructed project of specified quality at the lowest possible price.

Care should be taken when preparing public works bid documents to avoid language that is ambiguous or allows apparent interpretation. For example, the bid documents for a public works project in Minnesota included a statement that the owner had a right to "waive informalities or irregularities in a Bid received and to accept the Bid which, in the Owner's judgment, is in the Owner's own best interest." After the bids for the project were received and opened, the second-lowest bidder indicated that it made a human error on its bid and asked the owner to change the value of one of the bid alternates from a plus (+) sign to a negative (−) sign. The owner changed the sign under the pretense that this was a bid irregularity that it could waive. The second bidder then became the low bidder and was awarded the contract. The original low bidder filed a bid protest. Although the original low bidder was not awarded the contract, it did prevail in Minnesota Supreme Court, which stated that (Knoll, 1997): "A fundamental purpose of competitive bidding is to deprive or limit the discretion of contract-making officials in the areas which are susceptible to such abuses as fraud, favoritism, improvidence, and extravagance. Any competitive bidding procedure which defeats this fundamental purpose, even though it be set forth in the initial proposal to all bidders, invalidates the construction contract although subsequent events establish...that no actual fraud was present." Owners should be certain that their general provisions are adequately reviewed by appropriate legal professionals specializing in public works law. Owners should then be certain to scrutinize contract documents that it will use for bidding to ensure that ambiguous or contradictory language is not included in supplemental conditions or technical specifications.

2.3 Environmental Requirements

Federal, state, and local governments have many laws and policies to protect environmental quality. Most have a direct effect on a public works project, such as permitting requirements for work near waters of the United States, including wetlands. Construction projects that disturb as little as 1 acre of land may require mitigation of site drainage and runoff under the U.S. Environmental Protection Agency's National Pollutant Discharge Elimination System Construction General Permit (U.S. EPA, 2017). Others have indirect effects, such as pending regulations to reduce air emissions from nonroad diesel equipment (e.g., construction equipment). These environmental requirements can affect a public works project at all phases, including lead time to acquire permits, stop-work orders by agencies having jurisdiction for noncompliance with permit requirements, and postconstruction monitoring and reporting such as for invasive species of plants in surface restoration of wetlands.

2.4 Interagency Permit Requirements

Interagency permits include any necessary authorization to work in a public right-of-way, either to open-cut a road to install a casing pipe by a horizontal bore or for road patching, plumbing, or building (e.g., for lift stations). Collection system agencies may have blanket permits with other agencies (e.g., a state department of transportation) to allow for routine maintenance of facilities (e.g., cleaning catch basins). But these blanket permits may not apply to capital improvements, particularly when the collection system agency hires a contractor to perform the work. In some cases, the permitting agency may waive a fee if the permit is obtained on behalf of the collection system agency. Review by a historic preservation board or a parks commission also may apply. As with environmental permits, these interagency requirements can affect project schedule. Any necessary mitigation, such as improvements to a park or street, also can affect the project's cost.

2.5 Additional Agency Coordination

Additional coordination efforts may be required to permit and acquire rights-of-way such as railroads. These coordination efforts can sometimes be lengthy and require multiple iterations of design approvals, and therefore should begin early in the design process.

3.0 CAPITAL PROJECT DELIVERY METHODS

3.1 Design–Bid–Build

The traditional project delivery method for public works projects is design–bid–build (DBB). As the name implies, it is a sequential process. First, the owner prepares a set of contract documents, either with in-house resources or through a contracted design professional. Then the project is advertised, and bidders prepare estimates and submit bids based on the contract documents. The project is awarded, and construction of the project begins. The owner oversees construction either with in-house or contracted resources.

The product is well defined because all aspects are identified and designed before bidding. The DBB process is time-consuming, however, because of the sequential steps of the work. This method can limit the designer's understanding of construction labor, scheduling, material availability, and constructability. And it can create an adversarial relationship between the engineer, contractor, and owner. The nature of the process may cause the contractor to look for least-cost approaches to complete the project and meet or exceed its expected profit projections. This, in turn, creates the need for increased oversight and quality control by the owner.

3.2 Design–Build

Design–build replaces the traditional method of awarding separate contracts for design and construction. Design–build is a method of project delivery in which a single entity provides all of the services necessary to both design and construct the project. This single point of responsibility fundamentally distinguishes the design–build format from other forms of project delivery. A single source of responsibility reduces the owner's administration and management burden, protects the owner from liability, and reduces disputes concerning responsibility. Planning and procurement for successful design–build delivery differs in several ways from traditional project delivery. Strong owner involvement is required at the planning stage to determine the performance characteristics of the facility with sufficient clarity to establish fair competition. The procurement process also must allow designers sufficient flexibility to consider economical technical alternatives and require a guarantee for facility performance. It also must allow a basis for fair comparison between alternatives and provide a remedy when performance criteria are not met. Because teamwork is an important part of the process, owners must take great care to select qualified, functional design–build teams.

3.3 Construction Management at Risk

In construction management at risk (CMAR) delivery, the construction manager is expected to deliver a project to the owner within a guaranteed maximum price (GMP). The construction manager acts as an advisor to the owner during the planning and design phases and provides scheduling, budgeting, and constructability advice. During the construction phase, the construction manager is the equivalent of a general contractor. Because the construction manager is contractually responsible for a GMP, the nature of the relationship between the owner and construction manager is changed. The construction manager must protect their own interests in addition to those of the owner.

Construction management at risk is comparable to the traditional DBB method because the construction manager performs as the general contractor during construction. Specifically, the construction manager holds the risk of subcontracting the various tasks of the construction work to various trades and guarantees completion of the project for a predetermined, negotiated price following completion of the design. This delivery method is comparable to design–build because construction can begin before completion of the design. The construction manager can solicit bids and subcontract various portions of the work while design of unrelated portions is still being completed.

A disadvantage of this delivery method is the contractual relationship among the designer, construction manager, and owner as soon as construction starts. Once construction begins, the construction manager ceases to be an advisor to the owner and assumes the contractual role of general contractor. Quality of construction, issues associated with design, and schedule and budget can become points of contention. At this point, CMAR becomes analogous to the traditional DBB system, and adversarial relationships may result.

A significant benefit of this method is the opportunity to incorporate an experienced contractor's perspective into the decisions associated with planning and design of the project. Construction management at risk also allows for fast-tracking of early components before design is complete, saving time and associated costs.

3.4 Design–Build–Operate

Design–build–operate (DBO) project delivery entails the owner assigning a single contract to a firm that designs, builds, and operates a facility while maintaining legal ownership of the facility. The DBO contract gives the owner the flexibility of a single point of responsibility for a contract, which increases efficiency during the design and construction phase. The advantage of this single point of responsibility is realized during the operation of the facility, as the DBO contractor has a vested interest in ensuring a sound design and quality construction product.

4.0 OWNER'S REPRESENTATIVE

When the construction manager serves as a facility owner's primary representative for a project or program, that role is termed the owner's representative. The owner's representative is accountable to the owner for managing all or part of the planning, design, construction, and postconstruction phases. The construction manager protects the interests of the owner in its dealings with design and construction professionals and with other entities, both private and public. The owner's representative ensures optimum use of available funds, controls the scope of work, and is responsible for project scheduling. By coordinating the most advantageous use of the design and construction firm's skills and talents, the representative strives to avoid delays, changes, and disputes and to enhance quality. This method allows for optimum flexibility in contracting and procurement.

5.0 MAINTENANCE PROJECT DELIVERY METHODS

5.1 Force Account Construction

Construction projects involving repairs, small jobs, and emergency cases often are administered through a method called force account construction. Under this system, the owner acts as the prime contractor responsible for coordinating the project, and either subcontracts work to specialty contractors or elects to do the construction with existing in-house work crews.

5.2 Term Contracts

A term contract is a contract in which a source or sources of supply or services are established for a specific period of time at a predetermined unit price. For example, a utility repair term construction contract could consist of per-unit prices for the in-kind replacement of various sizes of gravity pipe. It might cover, for example, sizes of pipe ranging from 150 to 1050 mm (6 to 42 in.) at various ranges of depths such as 0 to 3 m (0 to 10 ft) and 3 to 5 m (10 to 16 ft), with a variety of surface conditions such as unpaved, light asphalt, heavy asphalt, or concrete pavement. As the utility project manager identifies individual defects in various pipe segments throughout the collection system, such as by closed-circuit television inspection, he or she can accumulate a list of required repairs. The project manager can then issue work orders to the contractor who holds the term construction contract to replace the sections of pipe with the defects using the established unit prices.

6.0 CONSTRUCTION CONTRACT DOCUMENTS

6.1 Constructability Review

As part of a quality review, the engineer should complete a constructability review of the construction documents, including bidding documents, supplemental conditions, technical specifications, and design drawings. It is important to consider the technical as well as the administrative components of the project during this quality review. The "Opinion of Probable Construction Costs" (OPCC) should be reviewed at various intervals during design. These reviews should be completed in conjunction with the constructability review to ensure the project is being designed to an established budget.

As outlined in Section 3.0, "Capital Project Delivery Methods," an advantage of the construction management at risk and design–build approaches is the innate contribution of a knowledgeable construction contractor. When using the traditional DBB, force account construction, or term construction approaches, the owner should have the design reviewed by an independent individual or firm with good working knowledge of the construction methods applicable to the project. The owner should be aware, however, that a contractor who performs a constructability review may become ineligible to bid on the project. Some jurisdictions consider the reviewing contractor to have an unfair advantage over other contractors.

6.2 Operations and Maintenance Review

The operations and maintenance (O&M) staff should have an opportunity to review and comment on the project at various stages. This real-world input by O&M staff provides the same benefits to the long-term viability of the project as contractor input provides to constructability. The O&M staff bear the burden of design and construction errors long after the engineer and contractor have completed the project. The O&M staff can assist with the development of life-cycle costs that can be used to evaluate capital improvement alternatives.

6.3 Value Engineering

Value engineering typically involves engaging a separate engineering group to examine a project plan (or conceptual design) to ascertain whether it will in fact deliver the value that the owner wants in a cost-effective manner and to consider whether alternatives will prove more reliable and cost effective. The process forces the owner to question and determine real needs. It also assists in identifying the main advantages and disadvantages of alternative engineering options. Properly conducted value engineering studies should end with general agreement on the best engineering design for a particular owner, based on whatever the owner values most.

Value engineering can be introduced during the construction phase of a project. Because it is more difficult and expensive to change plans after significant resources have been committed to an existing plan, value engineering has been found to deliver the most value for the least cost during the earliest stages of a project.

Many construction contracts with public agencies include value engineering incentive clauses. Value engineering uses the contractor's special knowledge to reduce the cost of a project without sacrificing quality or reliability. A clause of this type provides the contractor with an opportunity to suggest changes in the plans or specifications (such as changing from opencut excavation to trenchless technology) and share the resulting savings.

6.4 Allocation of Risk

Risk is the probability of the occurrence of a defined hazard (i.e., a particular event that can result in adverse effects if it occurs) and the magnitude of the consequences. Risk assessment is the estimation and evaluation of the risk (i.e., the magnitude of the consequences together with the probability of consequences). Risk identification is an awareness of those risks that could adversely affect the outcome of the venture or project. Risk management is the identification, measurement, and economic control of risks.

There are several factors that present risk to the owner, contractor, and engineer, including errors and omissions in design; environmental, health, and safety conditions; weather; quality of work; subcontractor performance; and inadequate project definition or organization. The utility project manager should discuss with its design professional or construction manager—or both—the project manager's tolerance for risk. This will help drive decisions on the level of effort for project components such as subsurface investigations, payment methods, and umbrella insurance policies.

6.4.1 Unforeseen Conditions

Unforeseen conditions are one area in which an owner can affect risk and cost. Wastewater projects are risky, and the potential for unforeseen conditions along any portion of the length of a project can have profound effects. If an owner is risk-adverse and states in its contract documents that (1) the owner accepts no responsibility for subsurface conditions, or (2) the contractor assumes all liability for all conditions found at the site whether or not these could have been reasonably anticipated, the owner will not only put a high degree of risk on the contractor, but will place a financial burden on itself. If the contractor perceives an undue risk, this will be reflected in its bid prices. The owner should take the necessary steps to give the contractor as much information as possible via a thorough subsurface program or should be willing to share the risk of truly unforeseen conditions. The allocation and management of risk will be reflected in the cost of construction. Appropriate contingencies should be built into the project's budget commensurate with the level of risk and unforeseen potential impacts on the project.

6.4.2 Dispute Resolution

The Construction Industry Institute states, "Disputes arise when the field project managers are unable to resolve a disagreement and others (e.g., project sponsors, advisors, neutral parties) become involved in an attempt to facilitate agreement" (CII, 1995). Disputes are identified as occurring on the job site and are a function of uncertainties surrounding the project, problems with process (e.g., incomplete scope definition), and issues involving people (e.g., poor interpersonal skills). Involvement of "others" not involved in day-to-day activities, yet share the goal of successfully completing the work, can be an effective method of resolving a dispute. Neutral parties or advisors can be individuals who are mutually selected by the owner and contractor before construction begins. This "dispute resolution board" is charged with rendering a decision based on the merits of each party's interpretation of the dispute and not based on an allegiance to a particular party. The dispute resolution board may meet several times over the project's duration to address routine disputes.

6.4.3 Conflict Resolution

If a dispute cannot be resolved on-site, the conflict can be addressed away from the project using techniques such as fact-based mediation, nonbinding or binding arbitration, a minitrial, or private judging. Whereas a dispute resolution board is part of construction phase, conflict resolution is typically an issue-specific technique.

6.4.4 Escrow Documents

On large, complex, or risky projects, such as a wastewater tunnel, the contractor's bid documents can be held in escrow by an independent third party. If a conflict arises in which the contractor is claiming extra costs, its bid documents can be reviewed to ensure that there was a legitimate omission or anticipation of certain conditions when the bid was being prepared. The placement of bid documents in escrow needs to be specifically identified in the contract documents.

7.0 FINANCIAL CONSIDERATIONS

7.1 Payment Methods for Construction Contracts

Although there are many different types of construction contracts, they may all be grouped within the following significant categories:

- Contracts for which the contractor is selected based on lowest responsive and responsible competitive bidder. This procedure is typical for public works contracts. Typically, these contracts will be lump-sum and unit-price contracts.

- Contracts resulting from direct owner–contractor negotiations. These can be determined on any mutually agreeable basis, such as lump-sum, unit-price, or cost-plus fee.

7.1.1 Lump-Sum Basis

Lump-sum contracts stipulate a fixed price for project construction. The lump sum reflects the total cost of the contract project. The owner benefits most from this type of contract because the total cost is known in advance. However, lump-sum contracts limit the scope of work because all quantities and complete plans and specifications must be accurately known when the bid is made. The contractor assumes full risk because the job must be completed even if the total cost of the work may be greater than the contract price. Lump-sum contracts are often used in wastewater collection system construction for lift stations and other well-defined structures.

7.1.2 Unit-Price Basis

Unit-price contracts offer advantages for projects involving quantities of work that cannot be accurately forecast. In this contract, the estimated quantities are compiled by the design engineer, and the contractor bids the unit costs for carrying out the proposed work according to the contract documents. The contractor is obligated to complete the work at the quoted unit price per listed item. Because payments to contractors are based on units of work actually performed and measured in the field, the exact cost of the construction is not known until the project is complete. As a result, unit-price contracts require complete plans and specifications. The contractor assumes less risk with unit-price contracts than with lump-sum contracts. Unit-price contracts often are used in wastewater collection system construction for pipelines and appurtenances.

7.1.3 Cost-Plus Basis

Cost-plus (or time and material) contracts are used when the owner wants a particular contractor to do the work, when the scope and construction techniques are not known, or when expedited project completion is necessary. In a cost-plus contract, the owner pays all the costs plus an agreed management fee; thus, the contractor takes the least risk. The owner and contractor typically negotiate the contract scope. Then, based on preliminary drawings and specifications, the contractor estimates the price. This type of contract often is used in force account work such as emergency repairs for wastewater collections systems.

Examples of items typically included in the management fee are the contractor's head office costs, rents, taxes, insurance, interest on financing for the project, head office supervision and control costs, and contractor's profit. Under this type of contract, the contractor has the least incentive to keep costs down, though variations of this contract have been developed to give the owner some cost control and provide incentives to the contractor to reduce costs. These variations include cost plus a percentage of cost, cost plus a fixed fee, cost plus a sliding fee, and guaranteed maximum cost-plus cost savings sharing incentives.

7.2 Funding Mechanisms

7.2.1 Municipal Bonds (Local Financing)

Municipal bonds are used to pay for public projects such as the construction or improvement of water and wastewater systems and other public works. There are different types of municipal bonds, including general obligation bonds, limited and special tax bonds, industrial revenue bonds, revenue bonds, housing bonds, moral obligation bonds, double barreled bonds, tax anticipation notes, bond anticipation notes, and revenue anticipation notes. The primary difference among these bonds is in how and when the issuer will repay the bonds and how it makes the interest payments. In the case of general obligation notes, the bond is simply backed by the "full faith and credit" of the issuer. That means the local or state government that has issued the bond can use just about any means available to guarantee payments, including raising taxes.

Other bonds are issued with specific provisions to raise taxes or create a new tax. These are known as limited or special tax bonds. If the project being undertaken will generate revenue from tolls, wastewater fees, water bills, or other services, they are known as revenue or industrial revenue bonds.

7.2.2 Federal and State Financing (Grants and Loans)

There are a variety of small-scale federal (e.g., U.S. Department of Agriculture, Rural Utilities Service) and state (e.g., North Carolina Clean Water Management Trust Fund) grant programs. The Water Pollution Control Revolving Fund, also known as the Clean Water State Revolving Fund, is the largest source of loans for both point source (wastewater and stormwater) and non-point source water pollution control projects (U.S. EPA, 1987). More information on project funding and financing options is provided in Chapter 8.

7.3 Cash Flow/Cost Control

Cash flow is important for the contractor and engineer, who rely on revenue to pay staff, buy equipment, and earn a profit. But cash-flow projections also are important for the owner who must borrow substantial amounts of money to fund a project. If the owner is undertaking a long-term program or multiyear project, it wants to limit borrowing to only that amount that is necessary to fund a particular portion of the project. In some cases, the owner will ask the contractor to provide cash-flow projections based on anticipated construction production rates and unit prices.

Cost control is important to all parties. Again, the contractor and engineer need to control costs to ensure a profitable job. The owner needs to track costs to ensure that it has sufficient funds to complete the project.

7.4 Construction Bonds

A bond is a legal document that binds a third party to the contract to provide assurance that the contractor will perform the agreed-upon service. Under the bond, a party known as the surety assumes liability for the failure and debt of another. Bonds typically required in construction include bid bonds and contract bonds (performance bonds and payment bonds). Bid bonds assure the owner that the bidder, if selected for award, will sign the contract. Contract bonds guarantee that the contractor will complete the work according to contract documents. If the contractor fails to fulfill the contractual obligations, the surety completes the contract and pays all the costs up to the face value of the bond. Standard bond forms are available from many different organizations, such as the CSI and EJCDC.

Construction contracts provide that the prime contractor warrants the work against faulty workmanship and materials for a defined period after project completion. One year is a typical warranty period, although periods of up to 5 years or more are typical on certain projects such as utility construction. During the warranty period, the contractor remedies construction defects at no cost to the owner. To ensure compliance with the warranty, a type of surety bond called a maintenance bond may be required. A maintenance bond requires the surety, if the contractor defaults, to correct defined defects in the contracted work that occur within the warranty period.

7.5 Construction Insurance

Utilities and states typically have specific insurance requirements in their regulations to protect the public; these are incorporated to the construction contract. For example, construction contracts typically require the contractor to provide coverage for workers compensation insurance, the contractor's public liability and property damage insurance, and the contractor's contingent liability insurance. Some contracts require the contractor to provide property insurance to protect the project and liability insurance to protect the owner. Many contracts require the contractor to protect the owner and design engineer by accepting any liability that may be incurred because of operations performed under the contract. Insurance certificates must often be submitted as proof of insurance.

8.0 MANAGING CONSTRUCTION CONTRACTS

8.1 Project Roles

It is important to define roles for the project to minimize confusion and maximize communication efficiencies. In most cases, each project will have three roles: an owner, an engineer, and a contractor. The role of the construction manager can be fulfilled by the owner or the engineer and, in some cases, the owner will contract with a third party to execute this role.

8.1.1 The Role of the Owner

The owner's role is primarily to provide the project to be built. The owner provides the land and easements on which the project is to be built and supplies available data related to the project such as explorations, testing, and potentially hazardous environmental conditions. Additionally, the owner is responsible for making payments to the contractor when payments are due per the agreed-upon schedule or payments.

8.1.2 The Role of the Engineer

The role of the engineer during construction can be multifaceted, but one primary responsibility is to act as the owner's representative during construction. During construction the engineer should make periodic visits to the project site to observe the progress and quality of the construction. The engineer is to determine, in general, if the work is proceeding in accordance with the contract documents based on observations made during these visits. The site visits are subject to the engineer's limitations of authority.

The engineer has the authority to reject work that is not in accordance with the contract documents. It has the responsibility to render decisions on whether the work conforms to the requirements of the contract documents and is acceptable based on these requirements.

8.1.3 The Role of the Contractor

The contractor is solely responsible for the means, methods, techniques, sequencing, and procedures for executing the work. In some cases, the contractor may be required to render design services if the contract documents require such activities or if such services are the result of the contractor's means and methods to do so. The contractor is required to supervise, inspect, and direct the performance of the work. It provides competent and qualified supervision and personnel to complete the work. The contractor is to provide services, materials, equipment, labor, transportation, construction equipment and machinery, tools, appliances, fuel, power, light, heat, telephone, water, sanitary facilities, temporary facilities, and other facilities and incidentals necessary for the performance, testing, startup, and completion of the work, whether these items are specifically called for in the contract documents. The contractor may retain subcontractors and suppliers that are acceptable to the owner for performing parts of the work and is solely responsible for the performance of its subcontractors and suppliers. The contractor is responsible for obtaining and purchasing construction permits and licenses required by the agreement and is also responsible for all taxes and duties arising out of the work. The contractor is required to comply with all applicable laws and regulations and for initiating, maintaining, and supervising safety precautions and programs relating to the work. This responsibility does not relieve subcontractors of their responsibility for the safety of persons or property in the performance of their work, nor for compliance with applicable safety laws and regulations. The contractor will warrant and guarantee to the owner that the work is in accordance with the contract documents and is not defective.

8.2 Intent of the Contract Documents

The requirements of each part of the contract documents are as binding as if required by all contract documents, complementary in nature. Typically, the contract documents do not describe all of the work and may require additional details to be provided by the contractor. The contractor should study the drawings and specifications and report any discrepancies discovered. If clarifications are required, the contractor or owner shall submit questions to the engineer for official response.

8.3 Progress of Work

A construction progress schedule should be generated and maintained by the contractor before the start of construction. However, this task can also be completed by the owner, engineer, or construction manager. The schedule must be developed with sufficient level of detail to adequately monitor and account for the planning, scheduling, and execution of work having time-dependent tasks with specific start dates and durations that must be followed to meet the planned completion date of the entire project. The schedule must be developed utilizing the critical path method, noting the shortest duration from the notice to proceed to substantial completion of the project. Float time can be identified within the schedule; however, the float in the project typically belongs to the project unless identified otherwise in the contract. The contractor should update the progress schedule monthly to track the project's progress.

Delays in the progress schedule will be evaluated on a case-by-case basis. The contractor is responsible for any disruptions that are caused within the control of the contractor. The contractor will be entitled to an equitable adjustment in the contract duration if the owner or its agents determine the impact is not the fault of the contractor or owner, such as force majeure. In requesting contract time extension, the contractor must demonstrate that the critical path of the project is affected by the number of days requested.

8.4 Progress Payments

Before construction, the contractor should submit a schedule of values for approval. The schedule of values must be broken down with a sufficient level of detail to best evaluate when progress payments are requested. Progress payments must be submitted and evaluated in accordance with the contract documents, typically on a monthly basis. The review of the pay application will depend on the type of contract: Lump-sum, unit-price, or time and material (cost-plus). Each review will require an independent evaluation of the quantities completed within the requested pay period. Typically, the pay request should be accompanied by an updated progress schedule, material tickets installed for the pay period, equipment tickets, affidavits, certified payroll, and any other information with which to substantiate the requested costs. A reduction in payment may be applied to the pay application for several reasons, such as deficient work not being corrected. If reductions in payment are being considered, it is good practice to engage senior team members for additional experience and guidance. Retainage is typically withheld from each pay application, which is the percentage or value typically established by state or federal law. The type and size of the project will guide the decision on how much retainage is withheld. It is good practice to gain support from the contractor's surety to ensure it agrees to the terms of the contract regarding payment and retainage. Final payments are made upon receipt of all close-out documentation provided by the contractor. The final payment should be accompanied by consent of the surety to release payments to the contractor, and affidavits of debts and claim on the project should have been paid and settled.

8.5 Changes in the Contract

Changes to the contract documents can be completed by three processes during construction: change orders, field orders, and work change directives. It is important to note that contractually, no other means of communication can change the contract, including request for information, conversations, or submittals. Changes to the contract that involve (1) the performance or acceptability of the work, (2) the design (as set forth in drawings, specifications, or otherwise), or (3) other design or technical matters, must be supported by the engineer's recommendation. The owner and the contractor may amend other terms and conditions of the contract without the recommendation of engineer. Change orders are any changes that affect the contract price or contract time. Change orders may also be used to pay for undisputed work performed by a work change directive, make final adjustments to alternates or allowances, make final quantity adjustments, and resolve any claims. A work change directive does not change the contract price or the contract times but is evidence that the parties expect that the modifications ordered or documented by a work change directive will be incorporated in a subsequently issued change order following negotiations on the contract price and contract times. Engineers may require minor changes in the work that do not change the contract price or contract times by issuing a field order through the construction manager. The construction manager may issue a field order for nontechnical, administrative issues. Changes in contract should be accompanied by a sufficient level of supporting data by which to be evaluated.

8.6 Claims

Claims are typically assertions or demands made by the contractor when (1) contesting an initial decision by the owner concerning the requirements of the contract documents or the acceptability of work under the contract documents; (2) contesting the owner's decision regarding a change proposal; (3) seeking resolution for a contractual issue that the owner has declined to address; or (4) seeking other relief with respect to the terms of the contract. The contractor should substantiate its claim with supporting documentation with a sufficient level of detail to be considered. The owner and contractor may mutually agree to mediate claims, and this should be addressed in the agreement between the owner and the contractor.

8.7 Quality

Quality assurance is the sum total of management actions, strategies, and reporting functions required to assure the project owner that a constructed facility satisfies contract requirements and appropriate technical standards. Quality control is the inspection and testing function required to determine objectively the standard to which the work is being performed and facilitates the reporting of results to management. Whereas quality assurance is a management tool, quality control is a production tool. Many owners will hire an engineer to develop and manage a quality assurance/quality control program. Some agencies have their own quality assurance/quality control programs, and hire inspection and testing agencies to carry them out. Others require the contractor to develop a quality control program, and then assign a person or a third party to provide quality assurance oversight. Elements of the quality control program include project meetings, review and approval of submittals, field inspections, materials sampling and testing, and review and approval of record drawings. Systematic documentation and filing are the most important parts of a quality assurance/quality control program.

As construction proceeds, testing of individual components is performed according to the manufacturer's recommendations and recognized industry practices to ensure that no damage will occur during large-scale testing. After construction is complete, total system testing, startup, and commissioning begins. System testing proceeds as required by the manufacturer's recommendations, engineering specifications, and recognized industry practices. Results of these tests are documented carefully and systematically. Adjustments and corrections that may require engineering approval are performed as needed and carefully documented.

After all systems have been tested independently, the complete facility is ready for startup. Startup is performed in accordance with an O&M manual, which incorporates all relevant data from equipment and instrumentation manufacturers and information provided by the engineer describing how the entire facility is designed to be operated and maintained. After startup, the facility is tested and calibrated for optimum performance and efficiency. In some cases, the engineer may order modifications to the facility to satisfy performance requirements or improve efficiency.

Following commissioning, a final inspection is made and job files, shop drawings, and O&M manuals (updated to include results of all facility calibration tests and any changes made during commissioning) are turned over to the owner.

8.8 Warranty/Guarantee Administration

A warranty is a guarantee that a facility will function according to stipulated, agreed-upon performance criteria for a stated period of time. Warranties also provide for a remedy when this does not occur. Typically, these require the contractor to return to a facility and fix or replace any item that is not functioning to the requirements of the contract documents. The contractor typically backs up the warranty with warranties from equipment suppliers. When a potential warranty issue has been identified, it is essential that facts be documented carefully to avoid acting on incorrect assumptions about who is responsible for a given situation.

9.0 SAFETY CONTROL METHODS AND PROCEDURES

The Occupational Safety and Health Act (OSHA) established a nationwide safety program for workplace safety and health (OSHA, 1970). Employers are required to promote a work environment free from recognized health and safety hazards and to meet OSHA requirements (OSHA, 2019). Construction contracts contain articles that require the contractor to conform to all applicable laws, ordinances, rules, and regulations related to safety. The contractor also must comply with any safety program policies instituted by public and local governments. Additional information on safety programs and policies related to wastewater collection systems is provided in Chapter 9.

10.0 UTILITY DAMAGE CONTROL (ONE-CALL SYSTEMS)

Underground utilities pose significant risks during construction because information about their location may be lacking. In 1996, the U.S. Department of Transportation Office of Pipeline Safety organized the Damage Prevention Quality Action Team to develop a national damage prevention campaign now known as Dig Safely. Each state has set up what is referred to as a one-call, or dig-safe system, so that one phone call locates all recorded utilities in the area of construction and notifies appropriate utility companies (Common Ground Alliance, n.d.). The APWA has developed *Recommended Marking Guidelines for Underground Utilities* (APWA, 2001). Uniform color-coded cards are available to standardize marking locations, changes in direction, and dead ends for buried lines. Proper utility location helps avoid accidental damage and service interruptions.

Subsurface utility engineering (SUE) is the process of locating and depicting the location of underground utilities according to an industry-established level of quality. This practice has been promoted by the Federal Highway Administration as a cost-effective means of providing accurate underground utility information for design, which otherwise may cause costly construction delays. For more information about SUE, refer to ASCE Standard 38-02 (ASCE, 2002).

11.0 REFERENCES

American Public Works Association. (2001). *Recommended marking guidelines for underground utilities.*

American Society of Civil Engineers. (2002). *Standard guideline for the collection and depiction of existing subsurface utility data (38-02).* https://sp360.asce.org/PersonifyEbusiness/Merchandise/Product-Details/productId/232103356

Common Ground Alliance. (n.d.). Promoting 811 and the safe digging process. Retrieved March 20, 2021, from https://commongroundalliance.com/811

Construction Industry Institute. (1995). Dispute prevention and resolution techniques in the construction industry. Research Summary 23-1.

Knoll, J. L. (1997). *The construction law briefing paper, 97-02: Lovering-Johnson, Inc v. City of Prior Lake*, 558 N.W. 2d 499 (Minn. Ct. App. 1997). Fabyanske, Westra & Hart, P.A.

Occupational Safety and Health Administration. (1970). *Occupation Safety and Health Act, Public Law 91-596.* http://www.osha.gov/pls/oshaweb/owadisp.show_document?p_id=2743&p_table=OSHACT

Occupational Safety and Health Administration. (2019). *Occupational Safety and Health Act* (Part 1926–Safety and Health Regulation for Construction). http://www.osha.gov/pls/oshaweb/owastand.display_standard_group?p_toc_level=1&p_part_number=1926

U.S. Environmental Protection Agency. (1987). *The Water Pollution Control-Revolving Fund* (Amendments to the Clean Water Act). https://www.epa.gov/cwsrf

U.S. Environmental Protection Agency. (2017). *National Pollutant Discharge Elimination System (NPDES) Stormwater Program's construction general permit.* https://www.epa.gov/npdes/stormwater-discharges-construction-activities

7

Public Policy and Community Relations

The goal of this chapter is to introduce wastewater collection system managers to the role of public policy and community relations in successful, sustainable system management. Every political subdivision, including city, county, state, district, or other authority, enacts laws, makes policies, and allocates resources to achieve specific ends. Public policy is the process of defining problems, building support, identifying solutions, implementing policies, and evaluating effectiveness of policies.

1.0 PUBLIC POLICY THEORY

Public policy is not a single, standalone decision by a governing body but a dynamic, interactive process that reflects the common mind of the public body. Public policy formation consists of five steps as shown in Figure 7.1. The time associated with each step can vary from weeks to years, depending on complexity of the problem and agreement within the public body.

Problem definition is the first phase. In this phase, a problem is identified with corresponding goals. Often, the issue is researched and studied to develop possible concept-level solutions. Agenda setting is the second phase, although some experts group problem definition and agenda setting into a single phase. In the agenda-setting phase, efforts are made by small groups with a special interest to raise awareness and support for the issue.

The policy-adoption phase consists of the development of possible alternatives and selection of a single solution or policy. This phase typically includes public discussions on the issue and may incorporate input from groups, such as cost estimation for the various alternatives identified.

The implementation phase puts the policy into action. Further decisions are made as theoretical approaches meet real-world situations. Those responsible for carrying out policy enact procedures. Depending on the nature of the policy, the implementation phase can include testing of the validity of the policy, such as through litigation.

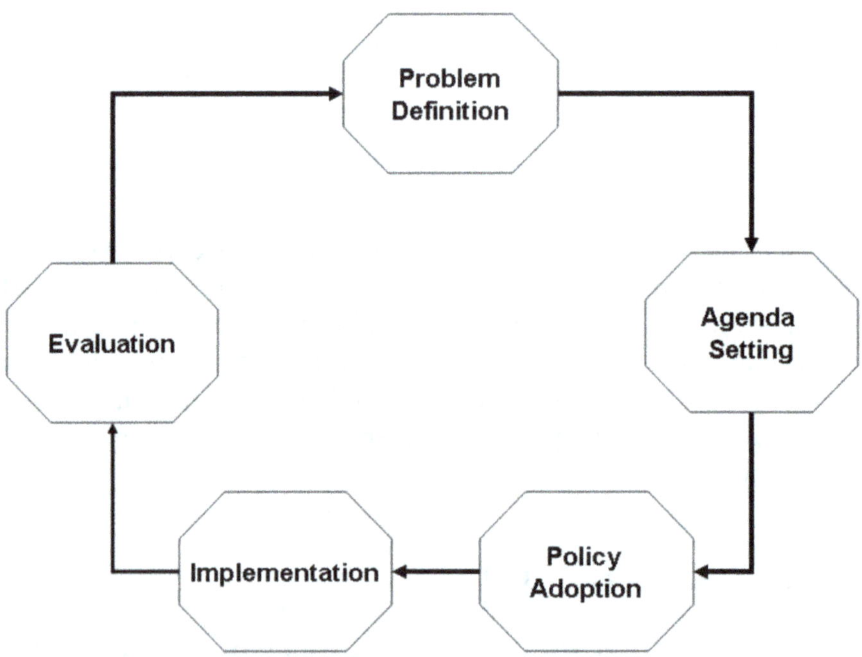

FIGURE 7.1 Five Steps of the Public Policy Cycle

The final phase is evaluation. After implementation is complete, the policy is reviewed to assess the degree to which the goals of problem definition and agenda setting were met. With gaps identified, the process begins again.

The issues that reach a governing body reflect the values and priorities of the community. Typically, scientific research alone does not play a significant role in agenda setting. Instead, the issues and causes that are embraced by the public find their way through the public policy process. As such, perception of a situation can carry as much weight or more than the reality of a situation.

2.0 PUBLIC POLICY IN WASTEWATER COLLECTION SYSTEM MANAGEMENT

Public policies affecting wastewater collection system management exist at every level of government. At the federal level, the Clean Water Act (CWA) first set the goals for clean water in 1972. Through the U.S. Environmental Protection Agency, policies were developed to effect those goals throughout the United States. Following the public policy cycle, the CWA has been amended several times to reflect the changing needs of society.

At the state level, additional regulations have been enacted to set goals to protect valued environmental assets. Although these regulations are coordinated with federal regulations, they frequently go further to provide additional or targeted protection. Similarly, regulations and policies are developed at the local level to further protect a locally valued environmental feature.

The purpose of defining all levels of public policy for wastewater collection system management is to set the baseline for characterizing current conditions, forecast those of the future, and manage the period in between. For any given condition, government can choose to intervene or to take no action. The decision of whether to intervene and how needs to be based on current conditions, such as land use, population, water use, and habitat, and how those conditions are likely to change. Because the environment is a dynamic, complex system, even small actions can have an effect, including unexpected, if not unwanted, side effects.

Characterizing the conditions that exist today can be done in many ways, ranging from occurrence documentation to technical analysis. The occurrence and effect of an ongoing problem can be documented by property owners, special interest groups, or wastewater collection system personnel using photographs and testimonials as evidence. Characterization using technical analysis may be done by in-house wastewater collection system staff or by a consultant. This analysis could include data collection, such as work order history, flow monitoring, and water quality sampling; laboratory analysis; the development and use of mathematical modeling tools; or some combination of these.

There are also many methods available for forecasting future conditions. Selection of a forecasting method should consider that environmental forecasting is often long-term; subsequent changes in conditions are likely. It should also be considered that environmental trends and action can interact to create new concerns. For example, correcting a manhole overflow may result in a basement backup. Wastewater collection, treatment, and discharge are one small segment of environmental management. Forecasting the effects of decision making can prove challenging when actions affect segments of the environment outside the normal limits of the wastewater industry.

3.0 ESTABLISHING WASTEWATER COLLECTION PUBLIC POLICY

As described earlier, in each municipality, county, district, or authority, the public puts forth issues and priorities that need to be diligently considered and incorporated into the allocation of funds, resources, and equipment. The methods that create public policy are unique to each community, reflecting individual histories and cultures. However, within the typical realm of a wastewater collection system, public policy topics include

- Pollution of waterbodies from the discharge of untreated wastewater;
- Property damage from basement or structure flooding;
- Nuisances, such as rats and cockroaches;
- Odors;
- Streambed erosion;
- Water reclamation and reuse;
- Private lateral maintenance;
- Design and construction of new and upgraded facilities; and
- Cost of services and the process for setting rates.

Other community-wide policy issues commonly affecting collection systems management decisions include affordability of rates and financing, asset investment, service extension, public–private partnerships, and economic development.

The wastewater industry often does not invest the time or resources to educate constituents or customers about the shared roles and responsibilities in managing use of the environment. As a result, the public may have performance expectations of the wastewater collector that are not possible or that are unrealistic based on the rates paid and the size of the wastewater collection system department.

Establishing local public policy recognizes and respects that the problems of the ratepayer are a problem of the community. For some issues, the policies established by the wastewater utility largely can address the problems. This is especially true when the wastewater utility owns and operates the system directly linked to the problem. For example, where odor is a problem around a pumping station, the wastewater utility has the facilities, personnel, and knowledge to install odor-control units or change the conditions to alleviate odor generation.

The utility also can establish policies that set boundaries between the responsibilities of the utility and an adjacent entity, industry, satellite community, or ratepayer. Basement flooding problems frequently fall into this category. Many wastewater utilities have limited the involvement of their personnel in flooding incidents to publicly owned portions of the system. Investigation of sewer laterals on private property is the responsibility of the homeowner. Such policies are often established for practical reasons, such as limitations of staff or equipment resources or to prevent public funds from benefiting private properties.

4.0 KEEPING PUBLIC POLICY RELEVANT

The key to keeping public policy relevant is to review the policy and update it when appropriate. The day-to-day responsibilities of the wastewater collection manager typically leave little time to review policy. Yet this review is an important part of having a collection system that utility staff and ratepayers consider successful. Policy reviews should be periodic but spaced far enough apart to determine whether it is achieving desired goals and, if not, where adjustments are needed.

A policy may have multiple review periods that should be specific to the policy. For example, a wastewater collection utility establishes an asset-management policy that does not include catastrophic infrastructure failures. One implementation goal is to perform small maintenance repairs to avoid large system failures. To meet this goal, the manager chooses to establish a program to televise the entire collection system once every 3 years. Under this example, the following may be an appropriate review cycle for the program:

- Every 6 months—review progress to determine whether the program of televising the entire system every 3 years is on schedule. Review staff hours and equipment usage.

- Annually—as part of budgeting for next fiscal year, review the program's operations and maintenance needs and recommendations for capital repair projects, such as lining. Review point repairs made and project funds needed for next year.
- After 3 years—upon completion of the first cycle, review use of staff and equipment, point repairs made, capital repair projects, and any emergency repairs or failures that occurred. Assess how well the procedure achieved the goal and make any adjustments. Assess whether the goal continues to be relevant. To the extent possible, coordinate with other actions under way to meet the policy goal, such as installation of capital assets.
- Years 4 and 5—review results of inspections for the two periods. Assess whether a 3-year period is appropriate, too short, or too long for individual sections of the system. Is the system condition better or worse than in the first cycle? If worse, are there specific causes of deterioration not understood or known when the policy implementation program was put in place?
- After 6 years—same as after 3 years; assess system wide whether the 3-year period is appropriate, too short, or too long. Assess how well the policy is achieving the goal. Assess whether the goal continues to be relevant.

Although it can be challenging to make the time to review policies, the evaluation phase is crucial to achieving the goal. Review periods should be tied to an event such as a specific maintenance activity, completing an inspection cycle, or a management activity, such as budgeting. Although significant time and effort are expended developing policies, there is no reason to expect that the first attempt at addressing a problem will find the perfect solution. The evaluation phase provides opportunity to make adjustments.

One component of the evaluation phase is to assess the goal itself and to determine if it is still relevant. A goal may be removed from public policy if (1) it is met and is no longer needed or (2) if the public no longer values the goal and is not willing to support it. In general, goals that are part of environmental decision making are goals that, once achieved, require some amount of continued, ongoing attention. An example of this is the significant effort under way to clean and protect the Chesapeake Bay. Once the bay has rebounded to the point desired, a policy will need to remain in place to protect the Chesapeake from degradation by future use.

The political will to retain a goal and public policy may wane if the public no longer sees a condition as a problem. Because public policy is rooted in the concerns of the public, it is subject, to some extent, to the cyclical nature of society. The wastewater collection manager can effect the sustainability of a policy by supporting sound policies and allowing others to fade. However, the purpose, intent, and goals of the policy should be continually publicized to garner future support.

5.0 PUBLIC POLICY AND COMMUNITY RELATIONS

Community relations are the link between ratepayers (the public) and policymakers. Through community relations, information moves both from the public to the policymakers and back from the policymakers to the public. Effective community relations are important to ensuring that ratepayers view the wastewater collection system as successful. The wastewater collection manager serves as the link between the public and policymakers. Community relations are a broad term that is used here to mean the system by which the wastewater collection system and ratepayers communicate. Terms such as public relations and public outreach are synonymous.

Each wastewater collection system has a way for ratepayers to indicate problems, express concerns, and share experiences. Frequently, this is limited to a customer service line that ratepayers can call when they have a problem. A few collection systems use a more proactive approach that strives to communicate with people when they are not in crisis, including the use of focus groups, public surveys, and community advisory teams. Working in community relations can require a lot of energy and a positive attitude. It is not a job everyone can do, but all staff should understand that they are the daily face of the organization. Ratepayers' perception of the utility will come from their experiences with the person on the phone and in the street—not necessarily the wastewater collection system manager.

As with technical goals, policies should be established based on managerial goals and address how the utility will interact with the public. The public is customer of the wastewater collection system, so it is important to understand their needs. Tools such as performance evaluations and surveys may provide valuable insight to their priorities and perceptions and enable a utility to better tailor services.

6.0 COMMUNITY RELATIONS AND PUBLIC EDUCATION

Public education is a facet of community relations through which information about the wastewater utility is disseminated. The purpose typically is to inform the public about activities in the utility but also can include soliciting input on specific topics. There are many formats available for public education:

- Mail:
  - Bill stuffers;
  - Newsletters, mailings; and
  - Flyers, handouts, door hangers, posters;
- News media:
  - Online news outlets;
  - Newspapers and magazines; and
  - Television or radio news;
- Digital channels:
  - The utility's own website;
  - Email blasts;
  - Social media platforms such as Facebook, Twitter, and Instagram; and
  - Neighborhood apps such as Nextdoor;
- In-person:
  - Presentations to community groups such as schools and homeowners' associations,
  - Booths at public events, and
  - Public meetings.

The Water Environment Federation's *Survival Guide: Public Communications for Water Professionals* (Wantland, 2002) helps wastewater collection managers develop strategies and tools to communicate with the public, interest groups, and the media. The Water Environment Federation (WEF) also has many brochures, bill stuffers, posters, and other public education materials available at www.wef.org.

Selecting a public education method should be based on (1) who the target audience is, (2) how they can most effectively be reached, and (3) what results are desired. For example, if a utility wants to educate people on a new interceptor project, the target audience may be identified as ratepayers in an effort to garner their support. In this situation, bill stuffers, newspaper articles, and participation at public events may be appropriate choices, whereas school presentations and flyers in apartment complexes likely will miss the target audience. The public education methods selected should also consider the demographics of the public. For example, print material using a larger font size will make reading easier for the elderly; non-English language copies could be available in select areas.

It is often said that wastewater is an "out of sight, out of mind" industry, but that everyone wants to ensure that they are getting the most for their dollars. Public education is the way a utility can promote accomplishments and explain problem-solving approaches.

7.0 COMMUNITY RELATIONS AND PUBLIC MEETINGS

The public meeting is another way to convey information on topics of critical concern. Of the public education methods listed in the previous section, special attention should be paid to public meetings because they provide an opportunity for face-to-face interaction between the wastewater collection system manager and the public. If not well-planned, however, public meetings can create more issues than they resolve.

Several guides and websites are available to help wastewater collection system managers, some of which are listed at the end of this chapter. The U.S. Fish and Wildlife Service, U.S. Department of Agriculture, Forest Service, and Oregon Department of Fish and Wildlife put together *The Public Meeting Survival Guide* (1990), which contains excellent advice on how to structure and prepare for public meetings. The following sections are based on information provided in this guide.

A public meeting may be appropriate when seeking information, advice, and involvement in making decisions, when clarification on an issue is needed, or when multiple groups are involved. The public will interpret a "meeting" to mean that the utility is willing and able to accept input. Alternate public education methods might be preferred in some situations, such as when a decision has already been made, when data is inadequate, when there is not enough time to manage the input, or when emotions are too high to allow constructive input. In certain situations after a decision has been made, however, it may be beneficial for the manager to hear stakeholder concerns and to explain why the work is needed or why the work needs to be performed as planned.

Once it has been decided that a public meeting will be held, the next question is how much involvement is desired or expected. A "presentation" or other public education method, such as newsletter or newspaper article, may be a better forum for sharing a decision while limiting involvement or input. Public meetings are interactive and allow all attendees to participate. They are informal and produce a general record of events. For some situations, a public hearing may be required. Hearings are formal and must meet legal requirements. Formal statements are sought and a verbatim record is generated.

In planning the public meeting, the manager should consider who the audience is. Questions to be considered include

- Who are they?
- What do they know about the topic?
- What is their attitude about the topic, key stakeholders, and the wastewater collection utility?
- Are all of the side issues represented?
- Are all affected groups represented?
- Who are the decision makers and will they be present?

Consider what type of meeting format will suit the goals and audience. Examples of meeting formats include open houses, small-group breakouts, panel discussions, question and answer, presentation and discussion, and explanation and testimony. Small-group breakouts are excellent for giving everyone an opportunity to voice opinions. One drawback is that two people may have a different experience based on who led the small group and comments were managed. Care should be taken in all formats to remain compassionate and open but not try to solve everyone's problems on the spot, which can be interpreted as dismissing the concern.

Time needs to be taken to properly plan the meeting. This process begins by identifying meeting objectives. One way to do this is to start a sentence with: "at the end of this meeting, I want the audience to." This sentence can help guide the rest of the planning. For public meetings, objectives should be sensible, measurable, attainable, relevant, and trackable. Managers should keep several suggestions in mind:

- Create an agenda that is followed;
- Ensure meetings begin on time;
- Include breaks to keep the audience comfortable and focused;
- Alternate speakers and activities every 10 to 15 minutes, if possible;
- Select methods for receiving input—brainstorm lists, written comments, verbal feedback, prioritized preferences, questionnaire, or email; and
- Provide information to take away, such as handouts or flyers.

Slideshows are a common and effective medium to use during public meetings. When developing slides, managers should consider the following guidelines:

- Set up and test equipment before the meeting begins;
- Include a maximum of 4 to 5 information items per slide;
- Use animation selectively; too much can be distracting for the audience;
- Use pictures and graphics to demonstrate points; and
- Discourage speakers from reading slides to the audience; rather, they should elaborate on information provided.

Also, speakers should avoid industry jargon, especially acronyms, that are not common within mainstream society. If it cannot be avoided, terms will need to be defined for the audience throughout the presentation. Otherwise, the presentation will sound as if it is in a foreign language and the information being conveyed will be lost.

Managers should be aware of any commitments made during the meeting to ensure follow-through. For example, if a utility representative promised that meeting minutes would be available in 24 hours and they were not, credibility will suffer. As an alternative, a timetable without exact dates can be provided with ways for people to submit follow-up comments or ideas.

It is important that utility representatives be honest and compassionate in public meetings to demonstrate respect and value for the role of everyone in the room.

8.0 COMMUNITY RELATIONS AND CUSTOMER SERVICE

For this discussion, customer service is the direct interaction between ratepayers and the staff of the wastewater collection utility. Commonly, customer service personnel manage day-to-day activities such as activating new accounts, closing accounts, accepting payments, answering questions, and responding to requests for service. Typically, customer service activities are performed by people who are dedicated to customer services but, especially in larger organizations, do not work with the engineers and operators who manage the system. Internal policies must be established to set a level of service that each customer can expect, such as the maximum time they must wait on hold.

A request for service is passed from the phone operator to the collection system operator for action. The collection system operators are in the challenging position of responding to a situation that has caused the customer some inconvenience or damage, such as basement flooding. Customers can be emotional—angry, distraught, confused—and expect the operator to help improve the situation. Although there are times when the operator can help, they frequently must tell customers that the problem is on private property and explain that policy limits what they can do. Collection system operators are trained in how to inspect, operate, and maintain a collection system, but typically not in how to interact with a person under duress. Many operators find this to be an especially burdensome and stressful part of their jobs. Significant time can be required to talk with the customer, affecting the overall productivity of the crew. Some communities have found success designating a person or persons to interact directly with the public about field work. This person allows crews to focus on technical performance of the job and provides the time and detail the customer needs. Whether this aspect of customer service is provided by the crew or a designated person, wastewater collection system managers must provide the interpersonal communications skills training needed to effectively perform the job. Poor customer service can make a difficult situation worse, taking up more staff time and resources and resulting in undeserved liability for the utility. Providing good customer service will ease the situation for customers and, in turn, improve the utility's standing within the community.

Internal policies also must be established for the performance of service request work. Policies for service requests aim to provide a high level of customer service and typically are based on the performance of work and maintenance of equipment. Examples of common policies include initially responding to a service request within 24 hours, closing out a service request within 7 days, issuing service performance questionnaires, and calling to follow up on completed service and equipment repairs within 10 days.

Customer service calls can be an indicator of the priorities of the public, which feed into the development of public policy. An electronic database should record calls, including the address of the caller, description of the complaint, and date of call. This provides a searchable data record that can be used to determine which parts of the service area have the most problems, whether problems are associated with a specific time of year, and what types of problems are prevalent, all of which can prioritize future activities. After the service request is resolved, this data set can be expanded to include cause of the problem, action taken, and man and equipment hours used. The collection and use of this type of data is discussed further in Chapter 3, Information Management.

9.0 COMMUNITY RELATIONS AND EMERGENCY PREPAREDNESS

In the event of an emergency, community relations will be the channels first tapped to convey information to the public. Since Hurricane Katrina and the disaster of the World Trade Center attack in New York on September 11, 2001, the nation has developed emergency preparedness guidelines to disseminate information and share resources (see Chapter 10). On a community-wide basis, the police or fire departments, or both, typically manage emergencies, often using a media spokesperson to speak to the public and press. However, in the event of a wastewater collection system specific emergency, such as a sewer collapse, the responsibility of communicating the extent of the situation to the public or to the police and fire departments may reside within the wastewater collection utility. Utilities should have in place plans for communications with customers and stakeholders during disaster conditions. This "crisis communication plan" should include drafts of messages that may be needed in a disaster. Wastewater collection managers are strongly encouraged to review the emergency management system in place in their community or region and consider how the utility can tap into the system in the event of an emergency.

10.0 REFERENCES

U.S. Fish and Wildlife Service, U.S. Department of Agriculture, Forest Service, & Oregon Department of Fish and Wildlife. (1990). *The public meeting survival guide.*

Wantland, S. (Ed.). (2002). *Survival guide: Public communications for water professionals.* Water Environment Federation.

11.0 SUGGESTED READINGS

Armstrong, J. S. (1999). Forecasting for environmental decision making. In *Tools to aid environmental decision making*. Springer-Verlag.

Barkenbus, J. (1998). *Expertise and the policy cycle.* Energy, Environment and Resources Center, University of Tennessee.

Dolan, R. J., Rose, T. D., Baker, R. A., & Barnes, M. J. (2004). *Managing the water and wastewater utility.* Water Environment Federation.

Innes, J. E. & Booher, D. E. (1999). Consensus building and complex adaptive systems. *Journal of the American Planning Association*, 65(4), 412.

International Association for Public Participation. http://www.iap2.org

Karkkainen, B. C., Fung, A., & Sabel, C. F. (2000, December). After backyard environmentalism. *The American Behavioral Scientist,* 692.

8

Budgeting and Financial Planning

1.0 INTRODUCTION

Budgeting and financial planning are important aspects of wastewater collection system management. Prudent financial management ensures that sufficient revenues are available to fund all necessary operations and maintenance (O&M), repair, and replacement programs. Financial management is necessary to maintain collection systems in good working condition, and as necessary, to generate sufficient revenues for system expansion to serve new customers.

This chapter will introduce the reader topics related to financial budgeting principles and practices:

- The operating and capital budgeting process;
- Economic analyses that can be used to determine the life-cycle costs of project alternatives;
- A recommended approach to setting budget priorities through asset management;
- Capital funding and financing alternatives commonly used to pay for wastewater collection system projects;
- Methods of determining revenue requirements and establishing user charges and service fees;
- The role of Government Accounting Standards Board, Statement Number 34 (GASB 34), in the budgeting and financial planning process; and
- Selling an asset-management approach.

2.0 OPERATIONS BUDGET

The operating budget is the primary financial plan for a utility. It guides management during the budget period (typically a fiscal year, but more frequently covering multiple years) for acquiring and applying resources to produce the targeted level of service. Development of an operating budget consists of three basic steps:

1. Preparation of the budget in a format that allows it to be reviewed and discussed at all organizational levels.
2. Review of the budget by approving authorities at various levels within the organization.

3. Presentation and adoption of the budget at the highest level of authority required to invest it with legitimacy.

Preparation of the budget begins with the definition of service level to be provided. This can become time consuming as organizations attempt to go through highly theoretical approaches to defining missions, goals, objectives, and desired results. However, there is a simple straight-forward approach: Start with the level of service that is currently being provided. This has the advantage of establishing a common ground of understanding throughout the organization. Although current level of service may not be formally established or defined as clearly as theoretically desirable, problems and successes may have emerged through the operation of the system in several forums: internal discussions and problem-solving sessions, employee/manager performance reviews, customer complaints and problem resolution, and meetings with top management and the governing body.

This current level of service can be implicitly accepted: "With this budget we will be able to continue at the same level during the coming year (or other budget period)." Or an attempt can be made to define what "current level" means in terms of organizational goals, objectives, and operational results. If so, these should be kept fairly simple because the budget must be clear and understandable at all levels of the organization and to the public. For example, level of service could be defined in terms of

- Uninterrupted conveyance of wastewater flow to the treatment facility;
- Wastewater flow to the treatment facility with less than two household backups per year;
- Response time to critical incidents; or
- System condition, expressed in terms of condition evaluation efforts.

Whatever service indicators are selected, the budget process should identify the level of service that is currently being provided.

The next step in the process is to value the current level of service in terms of the resources that are necessary to maintain the wastewater collection system and ongoing costs that are incurred. The cost of owning and operating a wastewater collection system typically include the following cost components:

- Personnel services costs (including salaries, wages, and employee benefit).

- Material and supplies costs.
- Machinery and equipment (minor capital) costs.
- Debt service costs (i.e., principal and interest payments associated with outstanding debt, if any).
- Taxes (such as property, gross receipts, or income taxes for investor-owned utilities; excise tax on gross receipts or payments in lieu of taxes for municipally owned utilities).
- Transfers (e.g., transfers to the general fund for services for municipally owned utilities, if any).

To develop a budget for the current level of service, a manager simply starts with the current period budget categorized as shown above. Within these categories, there will be more detailed line-items of expenditures, normally defined in the utility's chart of accounts. Also available will be actual expenditures to date for the current period as well as actual expenditures compared to budget for prior periods. This information provides a baseline for the manager to begin budget preparation.

The primary question is: "what resources are needed to continue the current level of service?" This is generally defined as the resources within the current budget. The next step is to value those resources in monetary terms for the prospective budget period. Generally, this is a simple process: Current levels of expenditure are increased by the expected level of inflation from the current period to the prospective period. Often, the utility's finance department or budget office will provide inflationary factors. In fact, it has become common for these groups to provide to operating managers with a forecast of the current budget compared to the prospective budget period, given assumed levels of inflation. If this is not the case, reasonable estimates of inflation should be developed by the manager.

The consumer price index published by the Bureau of Labor Statistics can be used as a guide for inflating personnel services and materials and supplies expenses (www.bls.gov). Another common guide for inflating costs is the Engineering News Record's construction cost index for a particular region. A more precise estimate might be obtained from the public service commission in the state, which often publishes an inflationary factor for use by regulated utilities. In addition, the utility's system for providing merit increases to employees should be considered and a reasonable estimate of those increases should be prepared. In an organization using collective bargaining, negotiated (or expected) labor contracts for salaries, wages, and benefits should override general inflationary factors. Also, contract agreements established with vendors for future costs should be taken into consideration when forecasting the cost of purchases of materials, supplies, and equipment.

The budgeted cost associated with paying debt service on existing, outstanding debt can be projected based on actual debt amortization schedules provided with bond and loan documents. Other costs, such as taxes and transfers, can be budgeted based on historical trends and discussions with general fund managers, agencies, or departments imposing such cost on the utility. The initial proposed budget for continuing the current level of service should, then, be composed of

- A statement of the current service level in as precise terms as possible, including the expected output of the organization.
- An expenditure budget by line item or object of expenditure, as defined in the organization's chart of accounts, which essentially consists of the current level of expenditure increased by expected price inflation to the prospective budget period.

This current service level budget can stand alone as the organization's budget proposal. However, it also can form a baseline for alternative budgets, which fall into two general categories:

- Service improvement options
- Service reduction options

To develop these alternative budgets, it is extremely important that current service level be defined clearly because the heart of the process is the ability to determine consequences on service level of alternative spending plans.

Service improvement options should be constructed to provide a clear connection between increased expenditures and improved service level. There also may be known cost increases, such as power costs for lift stations, increased fuel costs, or planned staffing increases because of new, previously approved infrastructure coming online. These types of increases may be necessary simply to maintain the current level of service and should be presented as such. If not approved, these cost increases would result in a lower level of service, which should be documented. The goal of providing a clear connection between expenditures and service level is best served by constructing these budgets in relatively small increments from the baseline (current service level), generally in the range of 3% to 10% overall increase. This allows reviewers to focus incrementally on costs and service level effects and to make an informed decision regarding suitability of the investment for the organization. Service improvement options should be constructed using the same format and line items as the baseline budget, so that they easily can be compared and evaluated.

Service reduction options are constructed exactly like service improvement options but represent reductions in expenditures and level of service from the baseline.

Each service change should be accompanied by a clear, precise statement of the effect of the increase or decrease in expenditures on service as represented by the baseline budget. Taken together, this constitutes a system of "target budgeting" that is highly manageable and transparent at all levels of the organization that have responsibility for development, review, and approval of the budget. A column format for the final document would allow easy comparison among options, as illustrated in Table 8.1.

3.0 CAPITAL IMPROVEMENT BUDGET

Utilities are capital-intensive by nature, and the development of clear, realistic plans for investment in capital assets is an essential part of utility management.

TABLE 8.1 Example Statement of the Effect on the Current Service Level of an Increase or Decrease in Expenditures as Represented by the Baseline Budget

Collection system budget options	Service reduction option 2	Service reduction option 1	Baseline (current service level)	Service improvement option 1	Service improvement option 2
Personnel costs	$ 8,550,000	$ 9,500,000	$ 10,000,000	$ 10,500,000	$ 10,000,000
Materials and supplies	$ 3,420,000	$ 3,800,000	$ 4,000,000	$ 4,500,000	$ 4,750,000
Equipment	$ 500,000	$ 950,000	$ 1,000,000	$ 1,350,000	$ 2,500,000
Debt service	$ 2,000,000	$ 2,000,000	$ 2,000,000	$ 2,250,000	$ 2,500,000
Taxes and transfers	$ 1,500,000	$ 1,500,000	$ 1,500,000	$ 1,500,000	$ 1,500,000
Total	$ 15,970,000	$ 17,750,000	$ 18,500,000	$ 20,100,000	$ 21,250,000
Change from baseline	−13.7%	−4.1%	0.0%	8.6%	14.9%

Description of service level options

Service reduction 2	Elimination of proactive inspection system implemented 2 years ago
Service reduction 1	Reduction of proactive inspections by 40%
Baseline	5% increase over prior year budget as a result of inflation and merit increases
Service improvement 1	Escalation of proactive inspection system to complete in 5 rather than 7 years
Service improvement 2	Improvements to lift stations, including telemetry

Capital budgets, which establish planned capital expenditures for a prospective budget period, should be based on sound long-term capital improvement programs (CIPs). In financial accounting terms, capital expenditures are distinguished from operating expenditures in that they are capitalized: the cost of a capital expenditure is entered into the accounting records as a fixed asset, then that cost is recognized over the estimated life of the asset through depreciation expense. This allows the cost of capital assets to be amortized over the life of the asset. Although this approach meets the objectives of financial accounting, budgeting must take into consideration the actual flow of financial resources needed to implement the capital program. Capital projects can be funded with cash generated through revenues of the system, or by issuing debt and paying principal and interest of the outstanding balance over a period of time consistent with the expected life of the assets. Therefore, construction and acquisition of capital assets creates the need for cash flow either to directly fund the projects or to pay the associated debt service. Cash flow requirements are the primary consideration in developing a capital budget.

This section addresses types of capital improvements, expenses, tools used to evaluate the cost and benefits of capital projects, and financing and funding of capital projects. A more comprehensive description of the capital improvement planning process can be found in Chapter 4.

A wastewater collection system capital budget should reflect two general classes of expenditures:

1. Renewal and replacement-costs that are incurred to extend the life of current assets through the rehabilitation or replacement of the existing asset.
2. Expansion or growth-costs that are incurred to expand the capacity of the system and meet the needs of new customers.

In addition, regulatory requirements can drive the need for some capital improvements. This need can be part of either replacement or expansion projects. For example, reduction of combined sewer overflows (CSOs) for water quality improvement may require replacement of an existing combined sewer transmission line. The replacement project may be sized to meet future demands, which means it will also have an expansion or growth element included. Identifying the purpose and drivers of capital projects is important in determining the financing options as described later in this chapter.

Capital expenditures should be categorized by class of expenditure, as noted above, and in terms of their function in the system. There are two major functional areas in a wastewater collection system:

1. Collection—small-diameter collector pipeline directly connected to user facilities and transmitting wastewater to the transmission system.

2. Transmission—large-diameter transmission pipelines and relating pumping facilities used to transmit wastewater from the collection system to the treatment facility.

A sound capital improvement program will be based on projections of future demand, regulatory requirements, asset system evaluation of useful life, condition assessments, and other factors that might influence the need for renewal, replacement, and expansion of facilities.

In composing a capital budget, a manager should consider the cost and benefits of capital project alternatives. In addition, the establishment of a capital budget should also refer to the long-term CIP. It is important to consider such questions as

- Have the factors and assumptions driving the CIP changed significantly since it was developed?
- Have events been accelerated or delayed requiring delay or expedition of projects in the CIP?
- If the CIP is based on specific funding sources, are those sources still available?
- Has the political climate changed significantly so that changes are required to the CIP?

This part of the process requires a review of the CIP for the short term. Long-term CIPs are developed based on estimates about the future that almost always require some degree of adjustment over time. Simply incorporating the CIP for a certain year into a short-term budget plan—for which funds will actually be appropriated through a legal process—could result in significant misapplication of funds.

Once the CIP has been updated to reflect current conditions, the capital budget for the prospective budget period is ready to be formulated. First, projects should be categorized to reflect the status of the projects in the budget:

- Projects continued from prior years
- New projects

A budget worksheet should be developed for each project, including at a minimum the following items:

- Project name and unique identification number;

- Expected initiation and completion date;
- Type of project or percent of type (renewal and replacement, expansion, or regulatory);
- Description of purpose and advantages of project;
- Effects if project is not completed should be quantified;
- Funds acquired in prior periods that are carried forward to fund the project, if any;
- Source of new funds to supplement existing funds to the extent necessary to fund the entire project; and
- Expected effect on the operating budget of the project, showing the amount of each year for at least 5 years.

Based on the project worksheets, a capital budget should be composed showing each project on a separate line, clearly identifying the type, funding sources, and priority. For the prospective budget period, the total of all project expenditures and funding sources summarized, showing the total operating and capital budgets for the period.

Capital funding sources are discussed later in this chapter. If a project is debt-financed, however, the amount of cash required to repay the principal and interest and to meet debt service coverage requirements during the prospective budget period should be reflected in the cost summary for the budget.

As with an operating budget, it is important to tie capital expenditures to the utility's level of service. A summary of advantages and disadvantages related to the level of service can be developed. The next section, which focuses on asset management, explains how to determine capital needs using prioritization that is based on a clear understanding of the consequences of a capital spending plan.

4.0 SETTING BUDGET PRIORITIES THROUGH ASSET MANAGEMENT

This section addresses strategies for determining the capital budget and forecast amounts for renewal and rehabilitation of wastewater collection system assets. A more comprehensive description of the CIP process can be found in Chapter 4. Although some rules for reinvestment are referenced—such as 2% of the asset value assuming a 50-year service life or funding a minimum of annual depreciation expense for renewal and replacement projects—other, more focused approaches exist, as discussed below.

4.1 The Problem

Wastewater utility managers often speak of being stuck "between a rock and a hard place." They have to address issues such as CSOs; sanitary sewer overflows; capacity, management, operations, and maintenance; and other state and federal requirements. Managers also are challenged with "build out," "no growth," "urban decay," and "price resistance." If a utility carefully and equitably funds and finances the cost of expanding system capacity, meeting growing demand is self-balancing as rate revenues and connection (capacity) charges flow in from new customers. However, the emphasis has shifted for many systems to rehabilitation and renewal of existing assets to maintain service levels for current customers and to remediate deficiencies under regulatory requirements. Utility managers will be faced with an increasingly difficult job to secure needed rate increases as utility infrastructure ages and replacement needs increase. Although this problem is not new, the necessity for action becomes more clearly defined with every passing year. This is because the easier, softer way (replacement deferral) has too often been taken to avoid the immediate rate effect. Renewal and replacement costs are increasing.

4.2 The Solution

These problems are not limited to older systems or to economically stagnant areas. It is simply a matter of time: if this is not happening, it will happen in the future unless action is taken. In the industry, the set of methods, concepts, techniques, and systems developed to deal with these issues often are collectively referred to as "asset management." Assets can be viewed as resources that are used over their useful life to provide services and produce revenue. Good asset management involves the acquisition, coordination, and application of all assets. A good asset-management system is characterized by

- Reasonably complete inventory of all system assets,
- Reasonably accurate condition assessment of all system assets,
- Estimated useful lives for all assets,
- Existing asset values,
- Replacement costs for assets,
- Projections of total replacement costs in future years, and
- Sustainable service level.

4.3 Underground Infrastructure: The Tough Asset

Underground infrastructure is only one several assets required to run a successful wastewater utility system. Infrastructure also includes treatment facilities; disposal systems; pumping stations; instrumentation; and transmission, collection, and distribution pipelines. However, buried infrastructure is out of sight, making it difficult to conduct condition assessments without considerable time and money and easy to overlook, which can result in neglect. Renewal and replacement of pipelines is a dirty, difficult job. It is also a thankless job because, as pointed out, it places a cost burden on the existing customers that is unmitigated by new customer revenues.

An effective asset-management process requires four determinations:

1. Condition
2. Criticality
3. Cost
4. Consequences

The first step is assessment of condition. Without condition assessment, the utility manager does not have enough information about the system. There are three basic approaches to maintenance of these assets:

1. Random response approach—fix it when it breaks.
2. Systematic naïve approach—because it all has to be fixed eventually and generally lasts 50 years, dig up 2% per year and replace it.
3. Systematic intelligent approach—invest money to assess asset condition and fix those needing repair.

Figure 8.1 shows the rate effects implied under each approach. The random response starts out good, but costs skyrocket as the system deteriorates. The systematic naïve approach, because it does not address the most failure-prone parts of the system first, produces significant rate effects in early years, and then levels out later on. The systematic intelligent approach, by focusing on assets in the worst condition first, produces a more level rate effect than the systematic naïve approach, but higher rates in the early years than the random response.

The second step, analysis of criticality, focuses the asset-management effort by determining the relative importance of each part of the system. A priority for repair and replacement can now be established by combining condition assessment (how likely is it that this asset will fail?) with criticality

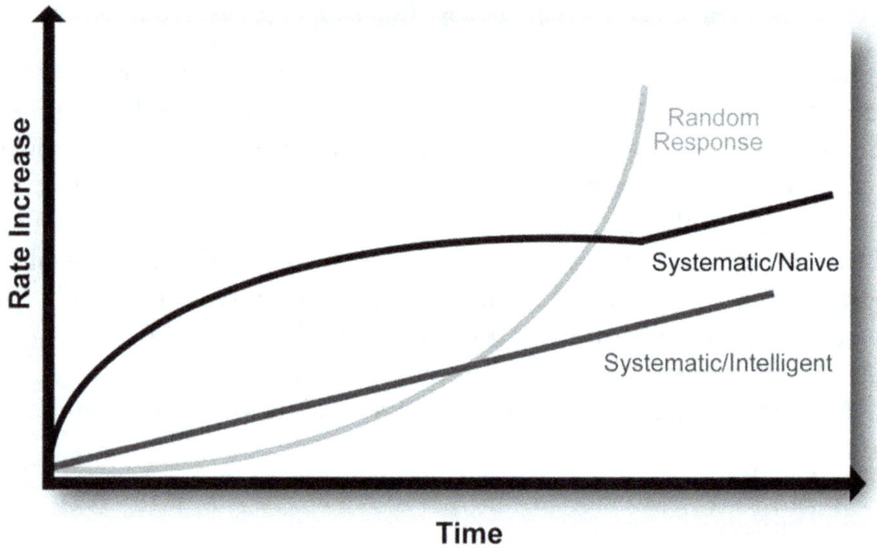

FIGURE 8.1 Rate Effects of Different Asset-Management Approaches

(if this asset fails, what will be the effect on the system in terms of liability, risk, and finances?).

The third step, cost determination and planning, involves estimating resources required to bring the assets to the desired level of condition and scheduling repair or replacement. This results in a prioritized CIP that provides an essential cost element for the forecast of rate effects.

The final step, evaluation of consequences, provides documentation of the probable effect on the system, to service and finances, of eliminating repair or replacement projects from the plan. The consequences of funding, or not funding the CIP provides the governing body with enough information to make decisions.

The conceptual asset-management process described above requires funding through the budget process; often, the budget can be viewed as a classic sales model. Management determines what the utility needs are and communicates that to the governing body. The governing body hears from customers and balances customer needs with realities of meeting regulations and properly maintaining level of service. However, customers may exhibit price resistance ("objections"). Management provides information to answer those objections, everyone discusses the issues, and the governing body makes a decision.

In a classic arms-length sales transaction, however, either party can simply walk away, which is not an option in this situation. Resolution must be reached regardless of whether the final decision produces a budget that meets the utility's needs. This makes development of sufficient information about the utility's needs, costs, and rate effect and persuasive presentation of this information essential to effective utility management.

4.4 Life-Cycle Cost Analysis

Life-cycle cost analysis is a useful tool that can aid in development of a capital budget. Life-cycle cost analysis can be used to help determine the total cost of implementing a capital project, compare costs of project alternatives, and make capital investment decisions, such as choosing between repair or replacement or other projects. Comparing the life-cycle costs of alternative capital investments is a way to evaluate their relative cost effectiveness and incorporate upfront capital costs, annual operating costs (or savings), and periodic replacement costs. Life-cycle cost analysis calculates the cost of an investment, including initial and annual O&M cost, over its life. It is especially useful in comparing the overall cost of projects that have different capital investment and O&M cost characteristics.

Under a typical life-cycle analysis, costs over the life of the project are converted to present value so that future costs can be compared on an equal basis. This "discounting" of future costs is done to reflect the time value of money (i.e., a dollar today is worth more than a dollar tomorrow). This is true because a dollar today can be invested and earn a return. Similarly, a dollar spent today costs more because of the opportunity lost from not investing and earning return. The formula for calculating present value of costs is presented below in Equation 8.1.

$$\text{Present Value} = \frac{\text{Future Value}}{1 + \text{Discount Factor}} \tag{8.1}$$

When calculating the present value of future costs, it is important to select an appropriate discount factor. The discount factor should represent the opportunity cost—the cost foregone by investing in the project today rather than placing the money in an interest-bearing investment of similar risk. The utility's investment earnings rate commonly is used because it is measurable and typically reflects a conservative investment policy that corresponds to the project risk. Other benchmarks, such as the Department of Interior Bureau of Reclamation's change in discount rate for water resources planning also has been used (Department of the Interior, Bureau of Reclamation, 2020).

A significant limitation of present value calculations is that they cannot be used to compare projects with different service lives. When comparing the life-cycle cost of mutually exclusive projects with significantly different service lives, it is important to make adjustments to account for their unequal lives. There are two commonly used adjustments to the life-cycle cost analysis that allows for a comparison of projects with unequal lives.

1. Equivalent annual cost approach
2. Common life approach

4.4.1 Equivalent Annual Cost Approach

Under the equivalent annual cost approach, the net present value of each project is calculated over its service life. The present value of each project is then converted into an equivalent annual payment, and the equivalent annual payments for each project are compared. The annualized cost is determined by equating the present value cost of the project and the present value of the annual payment (Brealey & Myers, 2000). This approach assumes that the project will be replaced with an equivalent project at the end of its life span.

4.4.2 Common-Life Approach

The common-life approach converts the projects that are being compared to the same life span by one of three methods: (1) replicating the shorter life project until its useful life span matches the longer life project; (2) replicating each project until a common life is found; or (3) completing the analysis over the life span of the shorter project and calculating and subtracting a residual value of the project with the larger life span from the present value cost.

Life-cycle cost analysis for wastewater collection system projects typically include the following costs and considerations: first cost, project life, O&M cost, and repair and rehabilitation costs.

- First cost. The first cost is simply the bid price for a project. This cost includes the construction estimate for labor, material, contingencies, supervision, and administration. Historical data may be used to determine an appropriate value for prebid evaluations. If a least-cost analysis is included as a basis for awarding a contract, the actual bid price would be used in the evaluation.
- Project life. Project life is the number of years that the project is anticipated to serve its intended purpose. For example, the project service life for culverts and sewers under primary and secondary roads may be between 50 and 100 years. A variety of sources are available to determine the life of wastewater collection system assets, including equipment manufacturers, public service commission allowable service lives, and empirical data from field testing.
- Operations and maintenance costs. These costs include the cost of operating and maintaining the project such as labor, materials and equipment, routine maintenance, and preventative maintenance costs.
- Repair and rehabilitation costs. Repair and rehabilitation costs may extend the life of wastewater collection system assets and can be considered in a life-cycle cost analysis.

Assumptions and estimates should be carefully documented, and the cost should be analyzed to ensure the best value.

When completing a life-cycle cost analysis, it is important to decide whether the analysis will consider cost in nominal or real terms. Costs stated in nominal terms take into account anticipated future trends because of price inflation. Life-cycle cost analyses are commonly completed in nominal terms (i.e., future costs are escalated for inflation). However, as long as inflation is handled consistently in the life-cycle cost analysis, either approach will result in the same conclusions. Regardless of which approach is taken, it is important to use a consistent approach to inflation of costs. If costs are escalated for inflation, all costs in the analysis should be escalated for inflation. In addition, an inflation-adjusted discount factor should be used (i.e., nominal discount factor). If costs are not escalated for inflation, a noninflation adjusted discount factor (i.e., real discount factor) should be used (Brealey & Myers, 2000).

4.4.3 Steps to Completing a Life-Cycle Cost Analysis

In summary, the following steps should be followed for completing a life-cycle cost analysis:

1. Identify the objectives of the project and identify alternative ways of accomplishing the objectives.
2. Estimate the cost of each alternative project including first costs, annual O&M costs, and anticipated repair and rehabilitation costs.
3. Decide whether the analysis will consider costs escalated for inflation and use a consistent approach for inflation of costs. If costs are escalated for inflation, ensure that all costs in the analysis are escalated for inflation and a nominal discount factor is used. If costs are not escalated for inflation, ensure that a real discount factor is used.
4. Estimate the life of each project alternative being considered. If repair and rehabilitation costs are anticipated, factor in the extension of the life of the asset because of the repair and rehabilitation efforts.
5. Determine whether the project lives being compared are significantly different. If they are significantly different, determine whether the common life or the equivalent annual cost approach will be used.
6. Calculate the life-cycle cost of each project using the net present value method, and make adjustments using the common life or equivalent annual cost approach as necessary.
7. Compare the life-cycle cost of each project alternative.

5.0 CAPITAL FUNDING AND FINANCING OPTIONS

Once a project alternative is selected, funding and financing options should be considered. When considering project funding and financing, it is important to refer to capital funding and financing policies that may provide guidelines for the utility. For example, a wastewater utility that plans to spend $1 million each year on collection system rehabilitation may be able to generate sufficient rate revenues to fund the program, rather than to use external financing, according to its financing policy. Alternatively, the policy may specify that a large, nonrecurring capital improvement, such as a pumping station, may need to be financed with bonds rather than funded on a pay-as-you-go basis with rate revenues. In addition, there also may be policies limiting the level of debt service allowed. It is important to carefully evaluate the various funding and financing alternatives within the policy framework and select an option that best meets the utility's needs and policies.

5.1 Capital Funding Sources

Capital funding sources are the sources of revenue that can be used to pay for capital projects and capital-related expenses (e.g., debt service). There are several capital funding sources that may be considered for wastewater collection system projects as summarized below.

5.1.1 User Charges

User charges are a source of funding both O&M and capital expenses. User charges may include any and all costs associated with the provision of wastewater service, including cash-funded capital and debt service (WEF, 2005). User charges can be used to fund capital projects on a pay-as-you-go basis, similar to O&M expenses, and to fund renewal and replacement projects.

5.1.2 Ad Valorem Taxes

An ad valorem tax is a tax that is based on the assessed value of the property rather than the flow or characteristics of the wastewater discharged. The advantages of the ad valorem tax are its assessment and administration simplicity, and its inherent ability to collect revenue from undeveloped properties in the service area that may eventually connect to the system (WEF, 2005). The disadvantages of using this funding source is that the assessment is based on property values and does not directly relate to wastewater flow and service characteristics. In addition, tax-exempt properties that may be connected to the wastewater collection system would be exempt from paying for wastewater service.

5.1.3 Special Assessments

Special assessments frequently are used where wastewater infrastructure is constructed for a specific purpose or for the benefit of a defined area, such as installing or replacing local sewer collection facilities (Water Environment Federation [WEF], 2005). Property owners may pay a special assessment based on a uniform charge per household, linear front footage, or other methods. A referendum is often required to establish a special assessment district.

5.1.4 Impact Fees

Impact fees—otherwise known as capacity fees, connection fees, and system development charges—are one-time fees typically collected from a builder, developer, or new homeowner for new or expanded service (WEF, 2005). In growing or expanding communities, the revenues collected from impact fees can be a significant component of the revenue stream of the utility. Impact fees may be used to pay for growth-related wastewater collection system capital projects or for debt service payments for growth-related projects. There are, however, many state-specific legal requirements that should be understood before using this funding source. One disadvantage of this funding source is that revenues collected are dependent upon growth occurring and can be variable. Another disadvantage is the potential delay between when the growth-related capital project is constructed and when the impact fees are collected.

5.1.5 Grants

The amount of grant funding that is available today to fund wastewater collection system projects is much less than the grant funding opportunities that existed during the U.S. Environmental Protection Agency grant program of the 1970s (WEF, 2005). However, several federal and state grant funding programs still exist that potentially can be tapped to fund wastewater collection system projects. These include grants available through the Community Development Block Grant Program sponsored by the U.S. Department of Housing and Urban Development, Water and Environmental Programs sponsored by the Rural Utilities Service of the U.S. Department of Agriculture, and Economic Development Administration Grants, among others. Grants have specific eligibility criteria and project requirements, including for funding cycles, schedules, wage rates, eligible cost, and in-kind matching and for conducting legal, financial, and administrative reviews.

The effect of these requirements and conditions should be carefully evaluated to determine project eligibility and suitability. With the diminished availability of grant funds, competition is steep, and it is often difficult to qualify for funding. A project has a better chance of securing a grant if it helps solve a documented water quality problem or health hazard or if the utility is located in an economically disadvantaged area, especially if utility charges are greater than 1% to 3% of the medium household income.

5.2 Capital Financing Sources

Capital financing sources commonly are used to amortize capital project costs over time. There are several capital financing sources for wastewater collection system projects as summarized below.

5.2.1 General Obligation Bonds

General obligation bonds are municipal bonds that are backed by a pledge of the full faith and credit of the issuer and can only be issued by governments that have the authority to tax. This is the strongest pledge the government can provide; therefore, general obligation bonds generally have the lowest cost of the various financing vehicles available to local governments (American Water Works Association [AWWA], 2008). However, because many states have limits on the amount of general obligation bond debt that can be issued, wastewater collection system projects that are targeted for this type of financing must compete with other general fund projects.

5.2.2 Revenue Bonds

Revenue bonds are government bonds that are backed by the specific revenues that are pledged in the bond documents. Revenue bonds are commonly used to finance capital projects that generate revenue and that are expected to be self-supporting, such as a wastewater utility. Most revenue bonds are issued with provisions that provides assurance to bondholders that their money will be repaid. For example, it may include a rate covenant specifying that user charges will be established to generate sufficient rate revenues to meet all operating expenses, debt service, and debt service coverage requirements. Debt service coverage is defined as annual net revenues divided by total annual debt service (principal and interest). The result is expressed as a percentage of 100% and greater or as a factor of 1.0 and greater for general obligation and revenue bonds, respectively. Revenue bonds typically do not require voter approval, and the issuer typically has some flexibility in establishing a repayment schedule that meets the cash flow characteristics and needs of the utility. However, revenue bonds may have higher interest rates than general obligation bonds and may have marketing difficulties, particularly if the utility does not have an established earnings potential or a history of financial operations (AWWA, 2008).

5.2.3 Governmental Loans

Government loans from federal and state agencies typically are available to finance wastewater collection system capital projects. The Clean Water Act of 1987 established a State Revolving Fund (SRF) program to provide loans to wastewater utilities for capital improvements. In addition, the Water Infrastructure Finance and Innovation Act of 2014 (WIFIA) established the WIFIA program, a federal credit program administered by U.S. Environmental Protection Agency (EPA) for eligible water and wastewater infrastructure projects (U.S. EPA, 2020). Likewise, the previously mentioned Water and Environmental Programs sponsored by the Rural Utilities Service of the U.S. Department of Agriculture also offer loans. These loan programs provide capital project financing at interest rates that typically are competitive with market rates. These programs often are competitive and dependent upon availability of funds. In addition, the timing and prioritization of the project will affect the suitability of these loans. Therefore, the specifics of each loan program should be assessed to determine the suitability for financing wastewater collection system capital projects.

There are other variations and derivatives of bonds and loans from federal, state, and private sources that may be available to finance wastewater collection system projects. It is recommended that the reader refer to the capital financing references found at the end of this chapter, as well as local municipal and utility associations that typically provide links to this type of information.

6.0 REVENUE REQUIREMENTS AND FEE SETTING

Generating sufficient revenues to fund wastewater utility programs to properly operate, maintain, rehabilitate, and replace wastewater collection system assets is an important aspect of wastewater collection system management. The following provides a summary of industry guidelines and common practices for establishing wastewater utility revenue requirements, user charges, and service fees.

6.1 Financial Condition

A wastewater system relies on the proper identification of annual revenue requirements to assess its financial condition and set user charges and service fees. By properly managing its annual revenue requirements and following industry guidelines for fee setting, a utility can maintain its financial condition. A financial management process can provide local officials with a high level of confidence that resources are being used properly. Financial management reports can provide local officials with the information they need to evaluate utility performance and set a course for future action. Through sound accounting procedures and periodic audits of financial records and procedures, assurance is gained that the resources of the utility are properly safeguarded. The following section presents recommended procedures for determining annual revenue requirements and setting defensible fees.

The financial management cycle for utilities is a process (see Figure 8.2) that has two major functions:

1. Financial analysis, measurement, and planning provide feedback into the knowledge base of the service cycle to give managers and decision makers an understanding of the financial consequences of their decisions, both historically and prospectively. This information also is used to provide representations to third parties for accountability and financing purposes.
2. The process of financial analysis, measurement, and planning produces information that allows management and governing bodies to price the utility's services to recover costs (and for investor-owned systems, a fair return) in a fair and equitable manner. This allows the utility to bill for its services, collect revenue, and use that revenue to acquire, maintain, and manage its assets. The real asset in this

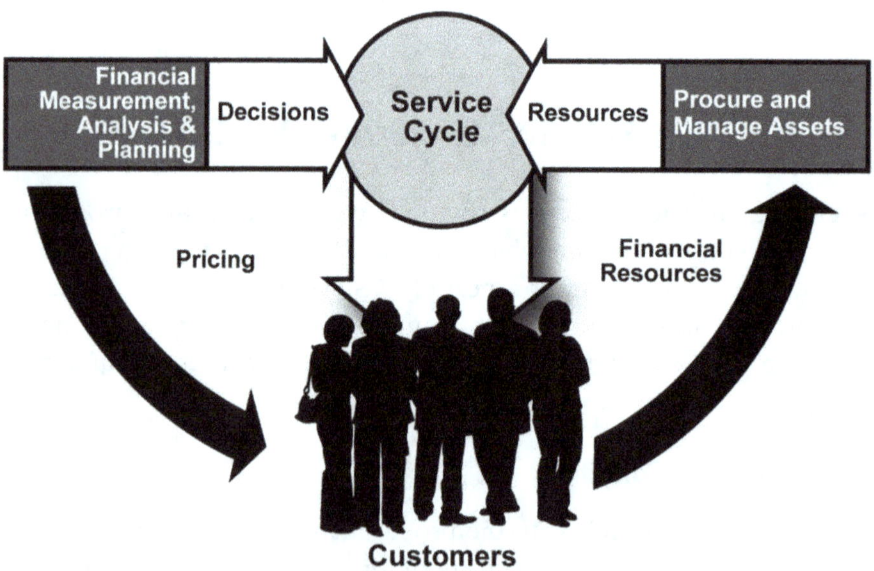

FIGURE 8.2 Financial Management Cycle

exchange is the customer base, which provides demand for level of service and revenue to pay for it.

6.2 Establishing the Basis for Fees and Charges

To maintain levels of service, a wastewater utility must establish the revenues it needs and the fees and charges necessary to generate it. Two approaches typically are accepted for projecting total revenue requirements: the cash-needs approach and the utility approach. A third approach is the utility approach with cash residual. Each has a proper place in utility practice, and each, when properly handled, can provide for sound utility rate (or fee) setting. A fully detailed discussion is *provided Financing and Charges for Wastewater Systems* (WEF, 2005). The following summarizes the approaches and identifies the cost components for each.

6.2.1 Cash-Needs Approach

Under the cash-needs approach, a wastewater utility's revenue requirements equal its total cash requirements for a typical year of normal operations, or test year. The cash-needs revenue requirement is the sum total of the utility's O&M expenses, debt service costs (principal and interest payments), payments in lieu of taxes, other applicable taxes, and any capital outlays of the utility that are not funded with debt, capital reserves, or other outside sources. Revenues, other than those derived directly from user charges, are subtracted from the revenue requirement to determine the user charge. This final figure is the portion of the total revenue requirements that the utility must recover through its user charges.

Because of its focus on actual cash expenditures during the test year, the cash-needs approach tends to work well with government budgets, making it the most common method of calculating revenue requirements among government-owned wastewater utilities.

To summarize, basic revenue requirement components of the cash-needs approach include

- Operations and maintenance expenses,
- Debt-service payments,
- Contributions to specified reserves,
- Cash-financed capital improvements, and
- Payments in lieu of taxes.

Depreciation is not included in the cash-needs approach. Expenses for O&M are projected based on actual expenditures and adjusted to reflect anticipated changes. Typically, O&M expenses include salaries and wages, fringe benefits, purchased power, other purchased services, rent, chemicals, other materials and supplies, small equipment that does not extend the useful life of major facilities, and general overhead.

The debt-service component of the cash-needs approach consists of principal and interest payments on bonds and loans. Capital expenditures are classified into three broad categories: normal accrual (routine) replacement of existing facilities; normal annual extensions and improvements; and major capital replacements and improvements. A utility should periodically review and update its needs in each of these areas to recognize changing conditions.

6.2.2 Utility Approach

The utility approach differs from the cash-needs approach in the way in which it considers capital costs. Instead of focusing on the cash outflows required to fund capital needs, the utility approach calculates capital costs using annual depreciation and a financial return that reflects the cost of capital on the adjusted used and useful utility plant in service, or rate base. To calculate the return component, the utility's cost of capital is multiplied by its rate base, which consists of the utility's used and useful net fixed assets and allowances for working capital and other adjustments (e.g., contributions in aid of construction).

The financial return component makes the utility approach the favored method of calculating revenue requirements for investor-owned utilities (and for those who regulate them). However, many government-owned utilities use the utility approach to calculate revenue requirements for extraterritorial customers. In these cases, a government utility may opt to use the cash-needs approach to calculate revenue requirements for customers inside its jurisdictional boundary and the utility approach for extraterritorial customers.

Utility-based revenue requirements may consist of

- Operations and maintenance expenses,
- Depreciation expense,
- Return on rate base, and
- Payments in lieu of taxes.

For a government-owned wastewater utility (such as a sewer district or conservancy district), the total level of annual revenue required may be the same under either the cash-needs approach or the utility approach. The O&M expense component of total revenue requirements typically is the same under both approaches. Under the utility approach, the annual requirement for capital-related costs consists of two components: depreciation expense and return on rate base.

Each state commission and regulatory body has its own rules, regulations, and policies for determining total revenue requirements. Therefore, when preparing for any rate matter, it is essential that the utility follow procedures and policies of the regulatory body in establishing its revenue requirements.

As discussed above, revenue requirements are the total operating and capital costs of the system typically identified for a single year. User charge revenue requirements are net of any revenues the system receives from other sources; those costs that the utility must recover from its user charges and service fees to balance its cash sources and uses for the test year.

6.2.3 Selecting a Test Year

A test year can be defined as the annualized period for which costs are to be analyzed and user charges established. According to the WEF (2005), the test year for government-owned utilities can include a historical year, a historical year adjusted for known and measurable changes, or a projected year. Use of a historical test year is appropriate when a utility is in a period of normal and sustainable operations without unusually high capital costs related to infrastructure expansion or replacement. For growing utility, use of a projected test year becomes more appropriate because revenue requirements will more accurately capture the anticipated increases associated with capital and operating costs.

6.2.4 User Charge Revenue Requirements

The portion of annual system revenue requirements to be recovered through user charges depends on a utility's financing policy and other sources of income. To determine the amount of revenue that user charges must generate annually, the total revenue requirements must be reduced by non-rate or other system revenues. Other system revenues are defined as all revenues except those derived from wastewater user charges. Some examples of non-rate revenues include

- Service fees;
- Penalties;
- Connection fees, tap fees, or system development charges; and
- Interest earnings.

Once the user charge revenue requirements of the system are determined, the costs can be allocated to system users. The costs to be allocated are referred to as costs of service as summarized in more detail by WEF (2005). Sound financial practices begin with determining the costs of serving the utility's customer classes. Communicating these costs to customers and designing user charges and service fees for recovery are the next steps to responsible financial management.

6.3 User Charge and Service Fee Types

Once rate revenue requirements are determined, wastewater user charges and service fees are established by utilities so that sufficient revenue is received from customers to ensure proper O&M, development, and perpetuation of the system and to maintain the ability to borrow. Rates, fees, and charges can be grouped into three broad categories:

- User charges or rates,
- Impact fees, and
- Other service fees.

6.3.1 User Charges

A wastewater utility's primary source of revenues is a user charge. User charges are the charges for service determined in a cost-of-service analysis. In a cost-of-service analysis, the cost responsibility of each utility customer class is determined through a cost-allocation process. Ideally, user charges will generate sufficient revenues to operate the system, preserve and build infrastructure, maintain financial integrity, and recover costs from customers equitably. Renewal and replacement of system assets is a critical component of revenues that need to be generated from user charges.

User charges take many forms depending on the community's goals and objectives. The basic components of user charges are fixed and volumetric charges. Fixed charges are charges per billing period that do not change with volume of wastewater flow. These charges often recover the costs of billing, general and administrative costs, a portion of debt service, and the costs of maintaining services.

Volumetric charges are those user charges that are based on volume of flow. These charges recover the remaining costs of service and are not limited to costs that vary according to flow. Volumetric charges can be uniform for all flows from all users or uniform by type or class of user (such as residential, commercial, restaurant, and hospital). Volumetric charges also can be tiered. Increasing volumes of flow can be charged at decreasing costs per unit (declining block rates) or at higher costs per unit (inclining block rates).

Finally, wastewater utilities that do not have volumetric charges that vary by customer type or class may assess "extra-strength" surcharges for industrial users. In this case, a customer's flow is monitored and compared to standard domestic strength wastewater. Surcharges typically are billed for each pound of constituent over a specified limit.

For a more detailed discussion on selecting user fee structures and developing rate schedules, see *Financing and Charges for Wastewater Systems* (WEF, 2005).

6.3.2 Impact Fees

Another significant funding source for wastewater utilities is impact fees. Impact fees, connection fees, system development charges, tap fees, facility investment fees, are all terms used to describe any fee or charge to recover capital improvement costs associated with system growth. These one-time charges are designed to recover the capital costs of growth from those causing the growth, rather than from the utility's existing customer base.

Without recovering investments in capital improvements for facilities expansion, the utility effectively would be subsidizing growth at the expense of existing ratepayers. For this reason, both existing and proposed investments in capacity are examined in calculating impact fees. The rationale for such fees is the unrecovered investment in available capacity, whether that capacity exists or is proposed.

In charging new customers for both past and new investments in capacity, the impact fee, like other such fees, promotes a concept in utility ratemaking called "intergenerational equity." Intergenerational equity means that existing customers do not subsidize new customers and vice versa. In many communities, this is often referred to as "growth pays for growth." Impact fees can be designed to minimize the subsidization of new growth.

Three approaches for calculating impact fees are commonly used for wastewater systems: buy-in approach, incremental cost approach, and combined approach (WEF, 2005). Each approach results in impact fees that are used to pay for growth-related facilities such as treatment plants, collection and interceptor mains, and lift stations. Revenues from these fees should be incorporated into the utility's financial management strategy.

6.3.3 Other Fees

To balance its sources and uses of funds, utilities often assess miscellaneous fees. Such fees range include basic administrative charges for service;

turn-on/turn-off fees; penalties for delinquent payments; and miscellaneous service charges, such as standby charges, availability of service fees, and installation or hookup fees. If a wastewater collection system has a reuse system component in its operations, charges for reuse or reclaimed water also may be appropriate. As in determining the user fees, establishing miscellaneous fees should be based on the costs of providing the service.

7.0 ROLE OF GOVERNMENT ACCOUNTING STANDARDS BOARD STATEMENT NUMBER 34 IN BUDGETING AND FINANCIAL PLANNING

As wastewater utilities establish user charges and service fees to support capital and O&M programs, they also will need to document financial results—an important aspect of the budgeting and financial planning process. This section reviews requirements of GASB 34. It discusses how budget priorities and asset-management programs are represented in financial reporting requirements for wastewater utilities.

The GASB is a private, nonprofit organization formed in 1984 to develop and improve accounting and reporting standards for state and local governments (www.gasb.org). Between 2001 and 2006, the GASB phased in the requirements of GASB 34, which was aimed at providing more comprehensive financial reporting information about the ability of governments to repay ongoing, increasing costs of infrastructure replacement debt and to continue to provide service. A discussion of GASB 34 requirements is pertinent to wastewater collection system management because it affects wastewater utility financial reporting requirements and provides an optional financial reporting approach that is based on asset-management practices described in other chapters of this manual.

The GASB 34 requires that all government agencies use one of two possible accounting approaches: (1) the conventional approach or (2) the modified approach. Many GASB 34 requirements are for general governments, such as cities and counties, which rely on general funds to support their asset-management activities. Before GASB 34, these governments did not list and depreciate infrastructure assets on their financial statements. Most wastewater agencies and governments with enterprise funds already listed assets on their balance sheets, so the conventional approach under GASB 34 imposed only minimal additional requirements on the utility.

Under the GASB 34 conventional approach, wastewater utilities need to satisfy the following requirements (www.gasb.org):

- Ensure that all infrastructure assets are included on the balance sheet and are depreciated using the accrual method of accounting.

- Include management's discussion and analysis as a narrative to the financial statements, which discloses significant financial activities, events, and conditions that affect financial position.
- Compare actual yearly expenditures to original budgets, not revised budgets.

These requirements are important to understand as capital and O&M programs are being developed and budgets are prepared. Necessary information can be gathered throughout the year to demonstrate results that programs and budgets were intended to achieve.

The modified approach to satisfying GASB 34 permits the current cost of maintaining assets to be reported on the financial statements in lieu of depreciation expense under several conditions. Under the modified approach wastewater utilities need to satisfy the following requirements (www.gasb.org):

- Maintain an up-to-date inventory of eligible infrastructure assets,
- Establish a target condition level for the assets,
- Complete a condition assessment at least every 3 years,
- Implement a formal asset-management system,
- Estimate and budget necessary funding to maintain assets at the target condition level,
- Conduct a review of the effectiveness of maintenance procedures,
- Ensure that assets are being maintained at or above the target condition levels, and
- Compare actual preservation expenses to budgeted preservation expenses.

If a wastewater utility complies with all of the requirements of the modified approach, it can report assets on a preservation basis, and may be able to reduce the amount of depreciation expense reported on its financial statements. This is important because depreciation expense represents a large, noncash cost to a utility that (1) reduces net income and (2) reduces the asset value on the balance sheet by the amount of the accumulated depreciation calculated because the asset was placed in service for utilities that use the accrual basis of accounting. By reducing depreciation expense, a utility will maintain a higher value of assets on their balance sheets, thus increasing net equity, and reducing the debt-to-equity ratio and reported leverage. As a result, the management practices that must be followed to use the modified approach could improve the reported financial condition of the utility. It also could improve the rating agency's view of the utility, although ratings are based on several factors.

Similarly, wastewater utilities that use the accrual basis of accounting—and to the extent that it has been expending funds to keep assets in good condition so that they are depreciating more slowly than measures of accounting depreciation would predict—the modified approach allows the utility to report the cost of maintaining the assets as an expense, rather than as an accrual of depreciation expense. In the short term, the modified approach could reduce utility expense and increase net income. Perhaps the most important benefit of the modified approach is that it requires a utility to implement "best practices" in the management of its infrastructure assets.

The additional cost required to implement the modified approach, however, is not insubstantial. For example, completing a condition assessment every 3 years and on assets tracking and updating condition would be costly but necessary under this approach. Regardless of which approach is used, however, management of assets on a proactive and preservation basis will provide long-term financial benefits. In addition, documentation of progress and results of capital and O&M programs is an important part of the asset-management process.

8.0 SELLING THE ASSET-MANAGEMENT APPROACH AND BUDGET PLAN

Utility management now has basic information to develop a long-term financial management plan. The key here is to "think long-term, act short-term." By putting the cost effects of the asset-management program into the context of comprehensive financial dynamics, a clear understanding of the future financial consequences of current management decisions emerges.

These steps create help inform decision making:

- Prepare a long-term forecast. Most utilities now use corporate planning models to develop financial plans. These range from simple spreadsheets to complicated commercial software packages. However, too often these are not incorporated into the budget process, which focuses on immediate needs and effect. It is essential that long-term forecasts be incorporated into the budget process to identify future "rate spikes" and plan to mitigate them over a longer period of time. This can include efficiency improvements, inflationary rate indexing to generate sufficient funds slowly over time, and changing capital improvements scheduling.
- Clearly demonstrate that costs drive revenue requirements. The forecast should identify the cost factors that create the need for additional revenue. In general, there are five major cost categories that should be displayed: operating costs (including taxes); cash flow requirements;

reserve requirements; capital improvement/debt service; and debt service coverage.

- Develop and demonstrate several scenarios. The financial planning process requires an understanding of the range of outcomes resulting from various economic conditions. Providing a worst case, optimal, and medium scenario can often facilitate the decision-making process. A range of assumptions regarding cost inflation, interest rates, financing terms, customer growth, asset failure rates, and other relevant factors should be explored. However, the factors adjusted within each scenario should be limited to keep comparisons clear and easy to compare and understand. Office productivity software, such as spreadsheets, can illustrate these factors easily and clearly. Graphics can help display the results in way that is easy to understand.
- Work interactively. A summary results page of the planning model can be projected onscreen at budget meetings, allowing scenarios to be explored interactively and alternative assumptions tested on the spot. This provides also provides a way for all parties to contribute to the effectiveness of the process.
- Keep the discussion in context. Whenever short-term solutions are suggested, both short-term and long-term effects on user charges should be shown.
- Be ready with consequences. The big question is always, "So what?" Seldom expressed so succinctly, it is nevertheless on everyone's mind and will be expressed in indirect ways. The development of consequences documentation is essential to providing a basis of confidence for the final budget decision.

Asset management provides an effective framework for setting budget priorities and addressing the challenges of infrastructure renewal and replacement. It is equally important that these determinations be integrated effectively into the budget process. By using an interactive approach incorporating long-term forecasting with short-term budgeting, a productive forum is created that helps to balance the needs of deteriorating infrastructure with rate effects. An understanding of the necessity of that balance can only be provided by a long-term perspective.

9.0 REFERENCES

American Water Works Association. (2008). *Water utility capital financing* (3rd ed., WEF Manual of Practice No. 29).

Brealey, R. A., & Myers, S. C. (2000). *Principles of corporate finance* (6th ed.). McGraw-Hill.

Bureau of Labor and Statistics. (n.d.). *Consumer price index*. http://www.bls.gov

Department of the Interior, Bureau of Reclamation. (2020, December 11). Change in discount rate for water resources planning. *Federal Register*. https://www.federalregister.gov/documents/2020/12/11/2020-27294/change-in-discount-rate-for-water-resources-planning

Government Accounting Standards Board. (n.d.). http://www.gasb.org

U.S. Environmental Protection Agency. (2020). *WIFIA program handbook*. (n.d.). https://www.epa.gov/wifia/wifia-program-handbook

Water Environment Federation. (2005). *Financing and charges for wastewater systems* (WEF Manual of Practice No. 27). McGraw-Hill.

9

Safety, Standard Procedures, Training, and Certifications

1.0 INTRODUCTION TO AND APPLICABILITY OF THE OCCUPATIONAL HEALTH AND SAFETY ADMINISTRATION

The Occupational Safety and Health Act of 1970 was enacted by Congress to ensure safe and healthful conditions for working men and women by authorizing enforcement of the standards developed under the Act; assisting and encouraging the States in their efforts to ensure safe and healthful working conditions; and providing research, information, education, and training in the field of occupational safety and health.

The Occupational Health and Safety Administration (OSHA) was assembled to carry out the Act. Many states have OSHA-approved state plans and, therefore, may have more stringent regulatory requirements that must minimally meet federal OSHA standards. Each employer is required to provide a safe workplace and job that is free from recognized hazards that are causing, or are likely to cause, death or serious physical harm. This requirement is sometimes referred to as the "General Duty Clause."

The OSHA regulations, contained in the Code of Federal Regulations (CFR), volume 29. Part 1903, Inspections, Citations and Proposed Penalties, and part 1904, Recording and Reporting Occupational Injury and Illnesses, apply to all employers with some exception for employers with 10 or fewer employees in a calendar year. There are two other parts within 29 CFR that are applicable to the wastewater collection systems industry: General Industry Standards, part 1910, and Construction Standards, part 1926. Any time a construction-based activity is being conducted, the construction standard will apply. Refer to OSHA applicability sections for further explanation.

2.0 APPLICABILITY OF HEALTH AND SAFETY REGULATIONS

Determining which regulatory requirements apply can be a daunting task and should be carried out by a designated safety professional that has had the appropriate training. Both the General Industry and Construction standards will apply to many different operations in wastewater collection systems. Each part of the regulation begins with an applicability section that will help the safety professional determine if the section applies. The Occupational Health and Safety Administration has programs such as consultative services that also may be used as a resource for determining applicability.

3.0 HEALTH AND SAFETY PROGRAM ELEMENTS

A health and safety program is a defined plan of action designed to protect employees in the workplace. The OSHA legislation requires that all workplaces implement some form of a health and safety program that meets or exceeds the minimum elements set forth in regulations. Minimum program elements may include (OSHA, 2007):

- Health and safety policy,
- Individual responsibility,
- Health and safety committee,
- Health and safety rules through written programs,

- Standard operating procedures (see Table 9.1),
- Employee orientation and training,
- Workplace inspections,
- Emergency procedures,
- Medical aid and first aid,
- Health and safety promotion, and
- Workplace-specific items.

Because each utility may differ in size, type of operations, and resources, one program cannot be developed for all wastewater collection systems. It is important that each utility develop a health and safety program that is specific to its needs. This chapter discusses general health and safety guidelines but should not be considered all inclusive.

3.1 Health and Safety Policy

In addition to a program, every organization should have a health and safety policy. The policy should be a statement of philosophy and guidelines to

TABLE 9.1 Sample Standard Operating Procedure Format

Your Company Name Here			**MANUAL**	
SUBJECT	**NUMBER**	**REV**	**EFFECTIVE DATE** 6/2005	**PAGE OF**
Name of SOP	**SUPERSEDES**		**PREPARED BY**	**APPROVED BY**
1.0 Purpose 2.0 Organization affected 3.0 Reference 4.0 Hazard description 5.0 Preventive measures 6.0 Engineering controls 7.0 Protective equipment 8.0 Personal hygiene and sanitation 9.0 Hazard communication and training				

which the organization and senior management are committed. The health and safety policy should receive the same level of enforcement and consideration as all other policies. Senior management is responsible for ensuring that the health and safety policy is carried out and enforced.

A health and safety policy does not have to be lengthy; in fact, it is better written clearly and concisely. The policy should contain a statement of management's commitment to the safety and health of employees, the organization's philosophy, and the general objectives of the program. It should also identify the person responsible for the program and state that all employees are responsible for adhering to safety rules and regulations. It is important to include a statement that poor safety performance will not be tolerated. The policy should be signed by the director of operations, kept up to date, and communicated to all employees.

3.2 Individual Responsibility

Individual responsibility for health and safety applies to everyone in the workplace, including top-level management. Each person is obligated to abide by established rules and regulations as they perform their job duties, which is why everyone must understand their responsibilities. Including specific health- and safety-related responsibilities in job descriptions can ensure that an individual understands what is expected.

Responsibilities for workers, first-line supervisors, management, safety coordinators, and the safety committee should be identified. Authority for enforcement and the level of enforcement also should be identified. This can be accomplished by developing an organizational flow chart for the program. Although responsibility and authority can be delegated to subordinates, management has ultimate accountability to ensure that the program is carried out.

3.3 Safety Committee

Establishing a safety committee is an integral step to creating an effective safety program. A safety committee encourages employee and management involvement and is a great tool for increasing awareness and imparting a sense of ownership. The safety committee should have an organized structure, a clear statement of purpose (committee charter), and established procedures for conducting regular meetings. Committee members can be appointed on a rotating schedule or can be made up of volunteers. Members should include management and all levels of workers. The safety committee also should be empowered through a clearly defined level of authority and resources to effect change. Some examples of duties that a safety committee may carry out are

- Reviewing and investigating accident reports, determining responsibility, and making recommendations for prevention;
- Reviewing statistics on lost time and no-lost time accidents, near-misses, first aid, and preventable and nonpreventable vehicle accidents;
- Periodically reviewing completed tailgate talks, safety meetings, and worksite inspections;
- Developing and reviewing job hazard assessments (JHAs) (see Table 9.2);
- Developing a training program;
- Reviewing written safety programs; and
- Reviewing operational changes for safety issues.

Attention to trends in the above statistics can be useful in promoting safety awareness at all levels. Essentially, the committee should be active in the development, implementation, and monitoring of all phases of the health and safety program.

3.4 Health and Safety Rules and Written Programs

As mentioned earlier, minimum health and safety regulations have been established by OSHA. States with OSHA-approved state plans may include regulations that are more stringent than the federal standard. Using these established regulatory requirements as minimum guidelines, a workplace must develop a written health and safety program. The workplace may create a written program that is at least as stringent as both the federal and state regulations. The written program must be specific to the industry and workplace and should include safe standard operating procedures (SOP) to achieve a safe and healthful workplace. Standard operating procedure development will be discussed later in this chapter.

Each organization should have developed written safety rules that are readily available and communicated to all employees. Many resources are available for assistance in developing these written programs. A list is included at the end of this chapter. Basic programs that typically are required in the wastewater collection system industry are listed below.

- Overall written safety program that establishes safety policy, organizational authority, and disciplinary action;

TABLE 9.2 Sample Job Hazard Analysis Form

Job title:	Job location:	Analyst:	Date:
Task #	Task description:		
Hazard type:	Hazard description:		
Consequence:	Hazard controls:		
Rationale or comments:			

- Excavation and trenching;
- Work zone safety;
- Permit-required confined space entry;
- Heavy equipment operation and rigging;
- Hazard communication;
- Energy isolation;
- Underground utility safety;
- Asbestos pipe;
- Personal protective equipment (PPE);
- Walking working surfaces/fall protection;
- Workplace exposure to wastewater;
- Respiratory protection program;
- Hearing conservation program;
- General safety awareness (nature exposures, heat/cold stress, etc.);
- Emergency action plan; and
- Machine guarding.

3.5 Standard Operating Procedures

An SOP should be created for each critical task using a process called a job hazard analysis (JHA), sometimes also referred to as job safety assessment (JSA). A JHA is the first step in creating an SOP. Employers should conduct JHAs for specific jobs by first breaking down each job into smaller individual steps. Each step of the task is evaluated to identify existing and potential hazards encountered or developed during the task. After risks have been identified, preventative measures are defined. The employee who carries out the task should be a participant in the JHA process because he or she is most familiar with the job and potential safety risks and can help develop realistic preventative or corrective measures.

Standard operating procedures are an important tool for all employees because they establish and ensure consistency and safety throughout a task and provide for quality assurance. An SOP should be used to train new employees and as a reference for infrequently performed tasks. An example of an SOP is included in Table 9.1. Some typical SOPs that should be developed in the wastewater collections system industry are listed below.

- Permit required for confined space entry,
- Inspection of high risk mains,
- Lockout/tagout,

- Rodding and jetting of mains,
- Electrical installations and safety,
- Work zone and traffic control,
- Easement maintenance and debris removal,
- Operation of heavy equipment, and
- Excavation and trenching (specific to a task; for example, installation of trench box).

3.6 Employee Orientation and Training

Health and safety education is an integral part of new employee orientation. Statistics show that inexperienced workers are likely to be involved in accidents more frequently than experienced staff. Generally, orientation will cover general policies, administrative items, and organizational relationships. Employee orientation should also include a distinct module related to safety such as emergency procedures; location of first-aid stations; location of health and safety rules and regulations; employee responsibilities; expectations in safety; and reporting accidents, near misses, injuries, and unsafe conditions. It also should cover use of PPE and the right to refuse to perform hazardous work if unsafe conditions exist.

General new employee orientation should not replace job and site-specific training. A new employee can be expected to absorb only a certain amount in the first few days, so it is beneficial to set up a "mentor" that the employee can go to with any questions. The department supervisor or safety coordinator should establish an in-depth safety orientation as part of job-specific training. This can include developing a safety training awareness schedule, reviewing applicable JHAs, SOPs, and PPE requirements, and establishing guidelines for on-the-job training. This level of orientation does not take the place of more in-depth training courses. The new employee should attend such courses as soon as possible.

In-depth safety training should follow orientation. An employee's failure to recognize hazards is one of the most frequent causes for serious injuries and illness in the industry. Hazard-specific training is required by OSHA as identified in the regulations and in written programs. Regulatory requirements mandate the type and frequency of training, which should be documented in the written program. Documentation of the written program, the training program, and training records are necessary to achieve regulatory compliance. The written programs and SOPs listed in this chapter are typical areas of concern in wastewater collections systems that require training (see also Sections 3.4, Health and Safety Rules and Written Programs and 3.5, Standard Operating Procedures). The lists are not meant to be all inclusive of training requirements; a safety professional should be consulted for further guidance.

The objective of a safety training program is to effectively raise awareness and skill levels to an acceptable level while easing the incorporation of policies and procedures into a specific job function. Different levels of training should be established based on job responsibility for managers, supervisors, trainers, and the worker. Developing a training matrix is an excellent way to track training requirements by position. Although training frequency varies, it should occur when there is a new employee, a transfer to a new job, changes to equipment or processes, or when inadequacies are identified. Training frequency should be identified in the written program.

Training can be accomplished through several avenues such as weekly "tailgate" meetings, regular monthly meetings, and specialized sessions. Weekly "tailgate" meetings are typically an informal training session held within a work group that covers a wide spectrum of safety topics and typically lasts approximately 10 minutes. Monthly training sessions may be delivered department wide and typically cover generalized topics, lasting approximately 20 to 30 minutes. Staff members who are knowledgeable on the subject, such as a supervisor or safety coordinator, may conduct the monthly sessions. Specialized training is generally more detailed and specific, can last from 1 hour to 1 week or more, and is instructed in a more formal setting by a qualified instructor. If the facility has an on-site staff safety officer, training coordinator, or other qualified individual, that individual can deliver the appropriate training. Training also can be conducted on- or off-site by an outside qualified vendor such as the National Safety Council or other industry organization. Written training modules should be developed for each topic being covered (other than for tailgate-style meetings). Modules include goals and objectives of the course, and items to be covered including applicable regulations, summary, and test.

A variety of training techniques should be incorporated into the session to encourage employee involvement and address different styles of learning. One example would be to use video clips and flip charts for discussion groups. It may also be required to incorporate a practical exercise such as with trenching and excavations or confined space entry. In this case, the instructor would arrange to carry out field exercises in which employees could participate in mock activities such as soil classification or entering a confined space. The employee should be able to demonstrate correct performance and knowledge of the topic by the end of the training session.

3.7 Workplace Inspections

Another key aspect in a safety program is regular, unannounced workplace safety inspections. Periodic inspections are a regulatory requirement that will help to identify and correct any existing or potential hazards or unsafe behaviors. For periodic inspections to be successful, targets should be set and monitored. Directors, managers, supervisors, and team leaders should be assigned a minimum number of safety inspections to complete throughout the year. Any variance from the target should be discussed at each individual's year-end performance evaluation. The workplace includes the main and remote operations locations and mobile job sites. Some items to include in inspections are work conditions and procedures; equipment, machinery, and tools; PPE; electrical safety; housekeeping; and adherence to SOPs and regulatory requirements.

Many other items may be included in the inspection and should be specific to the area or job being inspected. Any time that a danger to life or health is encountered, corrective action should be taken immediately. All inspection items should be documented, deficiencies noted, and corrective actions implemented. Inspections can be performed by a team or a designated individual that is knowledgeable of regulations and procedures (such as the safety committee or supervisor).

3.8 Emergency Medical and First-Aid Procedures

Emergency procedures are plans specific to the workplace for incidents such as fire, explosion, flooding, severe weather, chemical releases, or other events that may endanger personnel and require immediate response and decision making. The objective of the plan is to protect life and human health, protect critical infrastructure, and allow quick return to normal operations. Communication, training, and periodic drills are required to ensure effective implementation of the plan. Emergency preparedness and procedures are discussed further in Chapter 10.

3.9 Medical Aid and First Aid

Medical aid and first aid can apply in several different ways in the wastewater collection systems industry. Before starting work, employees are generally required to establish a health "baseline" by completing a general physical. More extensive medical testing may be required for certain job tasks such as the wearing of a respirator. If there are job exposures that are at or above established OSHA guidelines, employees must be enrolled in a medical monitoring program. A medical monitoring program may include annual hearing tests as part of the hearing conservation program, annual pulmonary function testing for respirator wearers, or annual blood work for possible chemical exposures. The safety coordinator or other designated person should work with the occupational health nurse to establish a medical monitoring program as applicable.

Second, procedures and policies for medical services should be established for employees involved in an accident or who have been injured. The policy should include procedures for reporting, obtaining medical care, and returning to work including provisions for light duty or modified duty. The OSHA standards require that medical services be available "in close proximity," which has been defined as 3 to 4 minutes. If an organization does not have access to medical services within these provisions, they must be provided. If an organization has designated first-aid providers, the written program should identify those responders, the location of first-aid stations, and guidelines for type and frequency of training.

Immunization programs, as recommended by the occupational health nurse or as required by OSHA for service operators, should be followed. Guidelines for employees exposed to wastewater in the eyes, mouth, ears, or through complete immersion should be accessible and well known to all employees. Health education must be provided to all employees. Shower facilities, special soaps, nail brushes, and other personal hygiene supplies also should be provided. The bloodborne pathogens standard covers wastewater workers if they are exposed to raw blood or other sources of infectious materials from hospitals or mortuaries. Employees exposed to sharp objects, such as needles, also are covered under the standard. Simple approaches, such as baseline health assessments before employment and uniform service with cleaning, are becoming standard practices in the wastewater industry.

3.10 Health and Safety Program Promotion

Much controversy exists regarding how health and safety programs are promoted. Many organizations have established safety incentive programs or recognition programs that award employees for good safety practices or for those who remain accident free. This causes concern among some who believe that this type of award or incentive program promotes underreporting of accidents and encourages the "walking wounded." Positive promotion without encouraging this type of activity can be accomplished through other tactics. For example, establishing a point system for those who attend the most safety meetings, include recommendations in near-miss reports, generate innovative safety ideas, or participate on safety committees are just a few methods that encourage proactive safety behavior without incentivizing underreporting. Promotion of the health and safety program should be well thought out and should include positive messaging that is fair and equal.

Health and safety minimum program elements are regulated and established by OSHA, although a state or organization may choose even more stringent rules. The elements discussed above are based on federal OSHA and are not intended to be all inclusive. More specific elements may be needed to address specific workplace activities. Utilities should consult with a safety professional or other agency such as OSHA's consultative services to ensure program compliance.

4.0 SAFETY PROGRAM ORGANIZATION

The safety program should be considered a key business element and have full support and funding from management. Program success ultimately is the responsibility of upper management. If the system is large enough, a safety coordinator should be appointed who directly reports to upper management and works with the department managers, front-line supervisors, and employees to implement the program. The safety coordinator should have the authority to reward employees or apply disciplinary and corrective action as needed with support from management. Often departments may not be large enough to have a full-time safety coordinator. In such cases, the responsibility ultimately falls on the front-line supervisors. Any person who is assigned responsibility for the program should have the support of management, including the time, training, and authority for safety.

5.0 SAFETY EQUIPMENT

Availability of approved safety equipment is necessary to maintain safety and health of employees. In the wastewater collection system, the tasks that are performed and associated hazards will dictate the equipment needed. Personal protective equipment and mechanical protective equipment are two types of safety equipment regularly encountered.

All PPE must be approved and rated for hazards encountered. Certain PPE must also be fitted to the person wearing or using the PPE. There are two agencies that typically approve PPE: American National Standards Institute (ANSI) and National Institute of Occupational Safety and Health (NIOSH). Safety glasses, steel-toe shoes, and hard hats are a few of the items that ANSI regulates. Other equipment includes eyewashes and showers, railings, and sound-level meters. NIOSH generally sets standards for items such as respirator cartridges, dust-mask-style respirators and other occupational exposure criteria, such as the permissible exposure level. Mechanical protective equipment standards for items such as machine guards, valve protection, some ventilation systems, and man-lifts also are established by ANSI.

A comprehensive assessment of facilities and field activities should be conducted by a qualified individual to determine what type of PPE or mechanical protective equipment may be required. Generally, all types of protective equipment should be regularly inspected and maintained. The OSHA has incorporated inspection requirements for equipment. If there is no regulatory guidance for inspections, maintenance, or replacement, the manufacturer's recommendations should be followed. A documented annual inventory and inspection is recommended at a minimum for all issued and stock PPE.

6.0 SUMMARY

Developing, implementing, and maintaining an effective health and safety program is a challenging task. The items discussed in this chapter are offered as guidelines as there is far too much information and specific applicability issues to provide a comprehensive outline. However, the sections addressed should provide a solid starting point to help in developing a thorough health and safety plan for a wastewater collection system. A strong safety program starts with management support and employee involvement.

7.0 REFERENCE

U.S. Department of Labor, Occupational Health and Safety Administration. (2007). *Safety and health management system* [eTool].

10

Emergency Preparedness and Security

1.0 INTRODUCTION AND OVERVIEW

Incidents in recent years have highlighted the potential effects of natural incidents and damage caused by humans to utility infrastructure, including wastewater collection systems. Although much of the focus has been on water systems, treatment facilities, and other infrastructure, wastewater collection systems also have proven to be vulnerable and their failure can result in significant consequences on public services and the environment. Regulatory changes also have increased the importance of preventing failures in wastewater collection systems. Wastewater collection system managers should put in place several measures to prepare for disaster incidents. Although some of these measures may apply more to larger wastewater collection systems, most of the guidance in this chapter is scalable to systems of all sizes. These measures are generalized in nature and intended to increase the resilience of wastewater collections systems to disaster and security risks but are not a replacement for system-specific risk assessments.

2.0 DISASTER AND SECURITY RISKS AND THEIR POTENTIAL EFFECTS ON WASTEWATER COLLECTION SYSTEMS

Potential incidents that can affect wastewater collection systems, for which wastewater collection system managers and operators must be prepared, can be grouped into three categories: natural disasters, incidents caused by humans, and technological failures. Each of these three categories is discussed in greater detail below.

2.1 Natural Disasters

Natural disasters that can affect wastewater collection systems include hurricanes, tornadoes, high-wind incidents, floods, flash-floods, landslides, ice storms, extreme winter weather, earthquakes, pandemics, and others.

Flooding from natural disasters has the potential to sever or wash out buried and aerial mains and other assets, to wash silt and other debris into manholes and mains, and to flood and damage pumping and lift stations. Winds, especially in conjunction with heavy rains, can cause mains and other assets to be damaged by uprooted trees. High winds and flooding have the potential to seriously damage and even destroy pumping and lift stations. In wind and flooding incidents, manholes and some other system components can be buried under silt and debris. Earthquakes pose a threat of breaks to mains and structural damage to other facilities.

All natural disasters have the potential to cause failures in the supply of electrical power to pumping and lift stations and communications systems and to disrupt supervisory control and data acquisitions (SCADA) systems.

All forms of natural disasters also have the potential to damage vehicles, heavy equipment, and other assets. Finally, many incidents can cause sanitary sewer overflows (SSOs) and combined sewer overflows, which can result in further damage to assets and to the environment.

2.2 Incidents Caused by Humans

Incidents caused by humans can be the result of unintentional or intentional acts of vandalism by disgruntled employees or other individuals.

Unintentional effects by humans on wastewater collection systems can include construction excavation and drilling damage to mains and laterals, accidental chemical spills into manholes and mains, vehicular accidents involving pumping and lift stations or other above-ground facilities, and employee errors.

Among the possible motives of individuals attempting to intentionally damage wastewater collection systems are to get attention among their peers, extract revenge for perceived wrongdoing, create an opportunity to be a hero, cause environmental damage, impair water supplies downstream, cause economic disruption, or attack a specific location through the wastewater collection system. Intentional damage caused by humans to wastewater collection systems can include main blockages and subsequent SSOs; damage to components, including mains, manholes, and pumping and lift stations; cyber-hacking or cyber-terrorism to shut down pumping and lift stations; contamination of the collection system; or use of some part of the system, such as chlorine, to cause injuries or damage.

Another recent example includes potential staffing shortages that could occur because of sickness. The ability of the utility to document standard operating procedures and have cross-training for multiple positions can mitigate the effect of staffing shortages.

2.3 Technological Failures

Both natural disasters and incidents caused by humans result in the technological failures listed above. Technological system components also are subject to intrinsic failures and loss of electrical power supplies, which can result in failures of pumping and lift stations, SCADA systems, and communications systems or breaks and structural failures of mains and other facilities.

All of the above types of incidents require that wastewater collection system managers conduct the planning and physical preparedness measures as discussed in the following sections.

3.0 PREPAREDNESS PLANNING ACTIVITIES FOR WASTEWATER COLLECTION SYSTEMS

Wastewater managers should implement several preparedness planning measures in anticipation of the effects discussed above:

- Develop an organizational culture of emergency preparedness and security,
- Prepare employees for disaster incidents,
- Conduct a vulnerability assessment to identify potential hazards,
- Develop viable emergency response plans (ERPs),
- Establish relationships with other local and state response agencies,
- Enter into mutual aid networks and agreements,
- Develop an understanding of the National Incident Management System (NIMS) and the Incident Command System (ICS),
- Be aware of cyber-vulnerabilities,
- Develop crisis communications plans,
- Conduct training and exercises, and
- Stay abreast of developments in the industry.

3.1 Develop an Organizational Culture of Emergency Preparedness and Security

The first step is to develop and promote awareness of what can happen in a system during and after a disaster. The preparedness process should include a worst-case scenario. As Hurricane Katrina demonstrated in New Orleans, Louisiana, in 2005, worst-case scenario planning should be pushed to the most extreme situation imaginable. Utility system managers should be careful not to overlook wastewater collection systems in their preparedness process. Utility systems should foster a security/disaster awareness culture among employees and provide training and reinforcement in various forms and at all levels.

3.2 Prepare Employees for Disaster Incidents

Many utility systems have learned that they also must address employee concerns when responding to a security incident or other disaster. Addressing these issues can determine the success of an incident response and benefit long-term employee relations.

Employees are a wastewater collection system manager's most valuable, and sometimes most vulnerable, asset in a disaster. Helping them prepare for disasters includes

- Educating them in the agency's plan;
- Assigning special disaster responsibilities, training, equipment and personal protective equipment;
- Apprising them of their role and what will be expected of them;
- Providing them with the necessary training, tools, and safety equipment;
- Establishing policies for mandatory work and criteria for leave;
- Providing shelter for employees with food, water, provisions for sleeping, and backup electrical power;
- Providing similar facilities and telephone hotlines for their families;
- Providing guidance for preparations at home; and
- Providing assistance in dealing with insurance and related issues associated with personal property damage.

Other measures to support employees include counseling as needed, paycheck cashing services, and post-disaster recognition and celebration, where applicable.

3.3 Develop Viable Emergency Response Plans

Water and wastewater utilities need to develop an ERP that incorporate all details necessary for effectiveness. Emergency operations plans and business continuity plans sometimes are used synonymously with ERPs.

The first step in developing an ERP is to identify the specific types of incidents that may affect a particular wastewater collection system or component. Emergency response planning requires detailed consideration of all potential disaster events. The plan should cover all employees and provide the level of detail necessary in a real event, including the availability of information on system infrastructure and minimum staffing requirements. Operational staff members need to be involved in assembling the details of the ERP to ensure their effectiveness.

Employees, representatives of other local agencies, and potential mutual aid providers are only as valuable as a collection system manager's ability to contact them during or following a disaster. Contact information can be the weak link in the disaster response chain. Managers must maintain this information in a non-electronic form that is current, accessible, and durable.

Emergency response plans should include

- Necessary internal and external contact information and details on the system infrastructure and where that information is accessible during and after disaster incidents,
- Provisions for materials and equipment that may be needed after a disaster,
- Maps and other critical system information, and
- Succession plans, in the event that key personnel are unavailable.

As Hurricane Katrina demonstrated, all local agencies must be prepared to provide for their own needs for at least 3 days following a disaster, before aid may be available from other agencies.

All personnel need to be trained in the details of their ERP. Emergency plans are not static documents and wastewater collection system managers must continue to monitor their vulnerabilities and refine and adjust their response plans as needed.

3.4 Establish Relationships With Other Local and State Response Agencies

Wastewater collection system managers should know the representatives of other local agencies. They also will need to provide training and information to staff to ensure coordination with staff at other agencies, such as

- Other sectors of the local water and wastewater systems;
- Local government management;
- Local emergency responders and emergency management agencies;
- Other local utilities, particularly electrical;
- State water and wastewater primacy agencies; and
- Local public health agencies and healthcare providers.

Recent disasters also have demonstrated that local agencies need to coordinate with and accept aid from nongovernmental organizations (NGOs), such as the Red Cross, and faith-based organizations (FBO). Although relationships with NGOs and FBOs will likely be more significant for other sectors of local government, collection system managers should be familiar with them. An excellent way to become involved with other local first responder agencies is through local emergency planning committees.

3.5 Develop an Understanding of the National Incident Management System and the Incident Command System

Related to the need to develop relationships with other local agencies is the need for wastewater collection system managers to become familiar with the NIMS and the ICS (which is a large component of NIMS), which includes

- Command and management structure and protocols (the ICS subset of NIMS),
- Preparedness,
- Resource typing and management,
- Communications (plain English),
- Supporting technologies,
- Maintenance, and
- Training.

Both the Department of Homeland Security and Federal Emergency Management Agency (FEMA) require NIMS compliance for fund eligibility.

3.6 Be Aware of Cyber-Vulnerabilities

Cyber-system vulnerabilities should be considered in security and disaster planning. Failure of SCADA systems can result in loss of system communication with and control of wastewater pumping and lift stations and other system components. Some of the best documented intentional cyberattacks on water and wastewater systems involve successful hacking to disable wastewater pumping and lift stations, causing SSOs, or inhibiting the ability of the public to communicate with the utility through webpages or email.

3.7 Enter Into Mutual Aid Networks and Agreements

Water and wastewater systems need to establish and develop mutual aid networks, both within and between states. Water and wastewater agencies in several states have been working to form such networks under the Water and Wastewater Agencies Response Network (WARN) and other models. The WARN establishes

- Mutual aid agreements,
- Procedures for communications and mutual aid activation, and
- Resource typing.

Mutual aid efforts in several states also have included:

- Establishment of a mutual aid coordinator's position in the state emergency operations center (EOC),
- Websites to post mutual aid needs, and
- Mutual aid responders' accommodations checklists.

A significant effort is under way by WARN and other organizations and agencies to develop an interstate mutual aid framework in response to frustrations in providing aid following the Florida hurricanes in 2004 and the Gulf Coast hurricanes in 2005.

3.8 Develop Crisis Communications Plans

Wastewater collection system managers need to prepare crisis communications plans (CCP) to support interactions with the public and the media. They need to develop news releases for various scenarios within their systems. Statements that may be needed should be anticipated and practiced. Core components of a CCP process include

- Developing the message,
- Establishing relationships with members of the media,
- Educating employees and other key stakeholders, and
- Identifying and training spokespersons and public information officers.

The U.S. Environmental Protection Agency (U.S. EPA) has been developing guidance on this task, which they refer to as "message mapping" (see https://cfpub.epa.gov/si/si_public_record_report.cfm?Lab=NRMRL&dirEntryId=156207). Although this need is more critical for water agencies, collection system managers also should be prepared.

3.9 Conduct Training and Exercises

Finally, wastewater collection systems should practice their disaster response plans. The U.S. Department of Homeland Security provides the prevailing guidance for exercise programs through its exercise evaluation program (HSEEP). The HSEEP identifies seven levels of exercises:

1. Seminar—informal discussion led by a presenter;
2. Workshop—formal discussion led by a facilitator;
3. Tabletop exercise—informal discussion of a hypothetical scenario;
4. Game—simulation of operations to depict a real-life incident;
5. Drill—activity to test a single, specific operation or function;
6. Functional exercise—examines coordination, command, and control between multiple-agency EOC; and
7. Full-scale exercise—physical, interagency response to a simulated event.

One of the best ways for wastewater collection systems to practice is through relatively easy and inexpensive tabletop exercises and games, which U.S. EPA refers to as enhanced tabletop exercises. Wastewater collection system managers need to conduct tabletop exercises with increasingly broad interagency involvement, both vertically and horizontally.

A tabletop exercise involves a simulated response to a hypothetical disaster, either natural or caused by humans. Unlike other types of exercises—such as full-scale exercises and drills, which often involve the mobilization of resources to a remote location and may even involve the use of actors as victims—tabletop exercises involve working through the disaster and response activities in one or more meeting rooms. Other than personnel and presentation and supporting materials, no resources are required. Advantages of tabletop exercises over full-scale exercises include lower cost, faster planning, scheduling and organization, and less vulnerability to the weather. Tabletop exercises also are sometimes called "sandbox" or "desktop" exercises. Most tabletop exercises are conducted in 6 hours or less.

Tabletop exercises lend themselves particularly well to many utility systems because incidents likely to occur involve relatively abstract components and situations, such as water and wastewater contaminants. This compares to incidents that other agencies may experience such as those involving hazardous materials, in which the location and components are specific, requiring other forms of exercises.

Wastewater collection system managers should conduct exercises or become involved in exercises for incidents that may already be occurring in the area. Exercises provide the opportunity to

- Test and practice ERPs,
- Get to know other local and emergency response agencies,

- Practice use of NIMS and ICS in interagency communications,
- Identify gaps in resources, and
- Test intra-agency and interagency communications.

3.10 Stay Abreast of Developments in the Industry

Finally, wastewater collection system managers should stay abreast of incidents and developments in the industry. Much can be learned from the lessons and improvements of others. This can be accomplished through reading industry literature, attending conferences, workshops, and seminars, and networking with peers.

4.0 PHYSICAL PREPAREDNESS FOR WASTEWATER COLLECTION SYSTEMS

Physical improvements that wastewater collection system managers should make or consider for disaster preparedness include

- Providing for emergency electrical power and/or portable pumping,
- Ensuring adequate protection of mains from waterways,
- Reinforcing aerial mains,
- Locating pumping and lift stations above flood elevations,
- Protecting vehicles and heavy equipment,
- Providing for necessary repair equipment and materials,
- Ensuring the availability of reliable communication systems, and
- Investing in a system hydraulic model to determine system capabilities and evaluate alternatives when needed.

4.1 Provide for Emergency Electrical Power

Hurricane Katrina and several other incidents have drawn attention to the dependency of water and wastewater systems, including wastewater pumping and lift stations, on electrical power. The provision for emergency electrical power when the power supply fails is one of the most important emergency planning actions of water and wastewater systems. Wastewater collection system managers should work toward having permanent, mobile, rental, or borrowed emergency electrical generators available to keep pumping and lift stations operational. Experiences in Hurricane Katrina demonstrated that trailer-mounted generators are more valuable for wastewater pumping and lift stations than skid-mounted units; the former can be moved from one pumping station to another. An on-site emergency power supply at each station is also a reliable solution. However, it should be noted that permanent on-site generators may also be affected during an emergency event and, therefore, all hazards should be considered when determining emergency power solutions.

Generators also require appropriate switchgear or generator connections and provisions for maintenance and fueling capabilities. Unlike most recommendations made so far, however, the acquisition of emergency electrical generators and associated switchgear is not inexpensive. Utilities that do not purchase generators should address this need through contracts to lease generators, provisions to borrow generators, or other arrangements. However, be aware, during widespread disasters, these generators will be in high demand and therefore may not be available when needed. The need for emergency electrical generators is a lesson that has been demonstrated repeatedly in disasters.

Emergency electrical generators are not resources that can be obtained and then ignored. Wastewater collection systems that have invested in generators or portable pumps must provide ongoing maintenance, testing, and capability for fueling the generators.

4.2 Ensure Adequate Protection of Mains From Waterways

Perhaps the most vulnerable components of wastewater collection systems are mains located along waterways. Because of the need to optimize gravity flow in collection systems, many mains are located along streams, rivers, and other waterways. Mains in these locations are particularly vulnerable to damage from flooding and erosion. Measures that wastewater collection system managers should take to protect mains from waterways include

- Locating the mains as far back from the waterways as possible;
- Installing manholes with rims above the flood elevation or with sealed manhole covers and vent stacks;
- Stabilizing the stream bank with rip rap stone or other measures;
- Providing regular easement maintenance to ensure access when needed; and
- Providing regular inspections of high-risk main locations, such as along waterways.

4.3 Reinforce Vulnerable Mains

The most vulnerable wastewater collection system mains are those that cross waterways, aerial mains in other low, above-ground areas, and those that are close to waterways. These mains are especially vulnerable to heavy waters, erosion, waterborne debris, and falling trees. Some wastewater collection systems have had success in reinforcing these mains by

- Using restrained joint pipe;
- Installing the carrier with spacers ("spiders") inside an outside-welded steel casing pipe;
- Minimizing the number of concrete piers to support the pipe;
- Increasing size, depth, and footing of piers;
- Improved straps and anchors to attach the pipe to piers;
- Extending double-walled pipe farther back from stream banks; and
- Structurally reinforcing manholes.

4.4 Locate Pumping and Lift Stations Above Flood Elevations

Stationary assets, such as pumping and lift stations, also should be protected as much as possible. This includes permanent improvements and temporary measures, such as

- Raising all facilities and equipment above the potential flood elevation;
- Raising particularly vulnerable components, such as electrical equipment, above the potential flood elevation; and
- Sandbagging of stop logging critical system components, like electrical components.

4.5 Protect Vehicles and Heavy Equipment

Collection system managers also must protect critical assets that are mobile. High-pressure jetting/vacuum trucks, for instance, are substantial investments for wastewater collection systems and are particularly valuable in disaster response. Wastewater collection system managers should do everything possible to put those and other critical pieces of equipment where they will be safe and available after an incident.

4.6 Provide for Necessary Repair Equipment and Materials

Wastewater collection system managers also should have on hand other resources as needed for follow-up to disasters. This may include

- Replacement pipe,
- Replacement and portable pumps,

- Repair parts for heavy equipment,
- Sandbags and other items, and
- Gasoline and diesel fuel for vehicles and equipment.

It is also critical to protect these assets from damage or loss in a disaster.

4.7 Ensure the Availability of Reliable Communication Systems

Finally, wastewater collection system managers need to ensure the availability of functional, reliable equipment for internal communications during and after a disaster. They should maintain their investment in traditional two-way radio communications that are not dependent upon the resources of others or subject to excessive public communications traffic. Recent advances in cellular technologies have led utilities to become increasingly reliant upon those technologies. However, they have proven to be vulnerable in disasters.

Traditional radio systems may require provisions for emergency electrical power to keep them operational in a disaster. Utility agencies should consider investing in satellite phones and participation in emergency phone systems, such as the Government Emergency Telecommunications Service.

Reliable internal communications systems should not be confused with interoperable systems between agencies. Although interoperable systems are valuable, internal communication is much more critical.

5.0 DISASTER RESPONSE IN WASTEWATER COLLECTION SYSTEMS

During and after disasters, wastewater collection system managers should undertake several measures to address possible security breaches and other emergencies in the systems, such as

- Activate ERP and open EOCs,
- Activate crisis communications plans,
- Check on the welfare of employees,
- Mobilize staff to critical remote locations,
- Conduct damage assessments,
- Activate emergency electrical power systems,
- Communicate and coordinate with other local and state response agencies using ICS,

- Activate mutual aid networks and agreements,
- Make emergency repairs, and
- Maintain records to maximize FEMA reimbursement.

5.1 Activate Emergency Response Plans and Open Emergency Operations Centers

The first thing that wastewater collection system managers should do during and after disasters and security breaches is to activate the ERPs that they have prepared and open their EOCs. This activation may include

- Contacting internal personnel and key outside agencies;
- Moving mobile and portable equipment to more secure locations;away from floodwaters and other hazards;
- Protecting critical facilities with sandbags or other measures;
- Protecting critical system information;
- Implementing succession plans or delegations of authority; and
- Establishing measures for employees to shelter-in-place at some work locations.

5.2 Activate Crisis Communications Plans

Wastewater collection system managers should use the crisis communications plans they have developed to share information with customers and stakeholders. This may include notifications of service interruptions, warnings regarding SSOs, or other key messages.

5.3 Check on the Welfare of Employees

As soon as a disaster has passed, wastewater collection system managers should strive to check on the welfare of employees and their families. Measures should be taken to provide assistance and enable the employees to participate in the agency's response to the disaster as quickly as possible.

5.4 Conduct Damage Assessments

Past disasters have demonstrated the need to make an initial assessment of the damage to a system before attempting to begin repairs. Damage to collection systems during extreme weather events often happen in areas that are not typically observed. Managers should ensure collection systems in these areas are investigated soon after the event. Damage assessments should check for several items:

- Interruptions of electrical power supplies to pumping and lift stations;
- Blockage of accessibility to facilities and areas of the collection system;
- Damage to pumping and lift stations;
- Debris over manholes;
- Damage to manholes and mains;
- Debris in manholes and mains;
- Damage to vehicles and heavy equipment; and
- Damage to supplies, materials, and other assets.

5.5 Communicate and Coordinate With Other Local and State Response Agencies

As the damage assessment is conducted, wastewater collection system managers should initiate communication and coordination with other local and state emergency response agencies, including

- Emergency management agencies for assistance that may need to be requested in anticipation of potential FEMA reimbursement,
- State regulatory agency regarding any environmental effects that may have occurred, and
- Local electrical utility regarding restoration of power.

5.6 Activate Mutual Aid Networks and Agreements

After checking on employees, assessing the damage, and determining any resulting needs, wastewater collection system managers should request assistance through the mutual aid networks they have established. It is important to use guidelines and resource typing to request aid using established protocols and terminology. Key needs and conditions that should be addressed for mutual aid responders include

- Housing and sanitation,
- Food and water,
- Employee work safety conditions,
- Communications systems,
- First aid and emergency medical services,
- Current inoculations required of responders,
- Safeguards for temperature ranges and weather conditions,
- Vehicles and equipment,

- Psychological support,
- Financial services, and
- Laundry services.

5.7 Make Emergency Repairs

As soon as the necessary resources can be assembled internally and through mutual aid, wastewater collection system managers should undertake necessary repairs to their system.

5.8 Maintain Records to Maximize Federal Emergency Management Agency Reimbursement

Finally, wastewater collection system managers typically will want to optimize receipt of potential reimbursement funds from FEMA. To do so, they should keep detailed records of their activities and expenses during the response.

6.0 SUGGESTED READINGS

American Society of Civil Engineers, American Waterworks Association, & Water Environment Federation. (2011). *Guidelines for the physical security of wastewater/stormwater utilities*. American Society of Civil Engineers.

Department of Homeland Security NRF Resource Center. (n.d.). http://www.fema.gov/emergency/nrf/

National Association of Clean Water Agencies (n.d.). *Decontamination Wastewater* [e-Library]. http://www.nacwa.org/index.php?option=com_content&task=view &id=47&Itemid&

U.S. Environmental Protection Agency (n.d.). *Drinking water and wastewater resilience*. https://www.epa.gov/waterresilience

Index

www.ingramcontent.com/pod-product-compliance
Lightning Source LLC
LaVergne TN
LVHW060558240526
839643LV00003B/40